Synthesis Lectures on Engineering, Science, and Technology

The focus of this series is general topics, and applications about, and for, engineers and scientists on a wide array of applications, methods and advances. Most titles cover subjects such as professional development, education, and study skills, as well as basic introductory undergraduate material and other topics appropriate for a broader and less technical audience.

Jay D. Humphrey · Jeffrey W. Holmes

Style and Ethics of Communication in Science and Engineering

Second Edition

 Springer

Jay D. Humphrey
Department of Biomedical Engineering
Yale University
New Haven, CT, USA

Jeffrey W. Holmes
School of Engineering
University of Alabama at Birmingham
Birmingham, AL, USA

ISSN 2690-0300   ISSN 2690-0327 (electronic)
Synthesis Lectures on Engineering, Science, and Technology
ISBN 978-3-031-39124-8   ISBN 978-3-031-39125-5 (eBook)
https://doi.org/10.1007/978-3-031-39125-5

This Springer imprint is published by the registered company Springer Nature Switzerland AG
The registered company address is: Gewerbestrasse 11, 6330 Cham, Switzerland

Preface to the Second Edition

We have been delighted to learn over the years, both from the publisher and many colleagues, that this book has been so well received by so many readers. Indeed, we have appreciated the many related invitations to present seminars, organize breakout sessions at technical conferences, and participate on national panels to convey the importance and methods of effective and ethical communication in science and engineering. We are thus pleased to present this Second Edition, which incorporates important updates and additions. Although many aspects of effective communication are timeless, there have been numerous changes in publishing and sponsored research; although the fundamentals of ethical conduct remain steadfast, many high-profile cases of misconduct continue. There is, therefore, a need for renewed attention to the *Style and Ethics of Communication in Science and Engineering* to present and promote advances that ultimately promise to improve the human condition.

In particular, in this Second Edition, we have built upon that which was presented previously, but have revised and expanded the content, especially in Chaps. 3 "Scientific Publications", 4 "Proposals and Grant Applications", 5 "Oral Communication and Posters", and 7 "Recordkeeping, Reproducibility, and Responsibility" to address important changes that have emerged since publication of the First Edition in 2009. There have been many changes in scientific publishing, especially to increase rigor and reproducibility, to increase data and software availability, and to decrease bias. There have similarly been many changes in how grant applications are reviewed, which has been especially challenging during periods of reduced funding. The COVID-19 pandemic changed the way science and engineering can be taught and communicated, particularly with Internet-based tools such as Microsoft Teams, Webex, and Zoom, and we have all had to adjust our style of presentation accordingly. Many conferences continue to emphasize the use of posters, which are often introduced via 2-minute oral presentations during lightning sessions but ultimately bring the presenter and listener together to facilitate more personal exchanges. Finally, with the availability of massive cloud-based digital storage, many journals and organizations require that data and software be widely available, which cannot be an after-thought as we prepare our work for presentation and publication. These and many other changes are addressed within this updated Second Edition, which also includes more Exercises to motivate and guide learning.

As with the First Edition, the goal of this book is to motivate the reader to develop an effective, individual style of communicating and a personal commitment to integrity while appreciating and preparing for the changing landscape of publishing, proposing, and presenting. We again wish the reader every success in these endeavors.

New Haven, CT, USA   Jay D. Humphrey

Birmingham, AL, USA   Jeffrey W. Holmes

Preface to the First Edition

How to write well.
How to publish your results.
How to secure funding.
How to speak well.
How to ensure integrity.

This book was written to help address these important aspects of beginning a career in science and engineering. In essence, scientists and engineers seek to discover and disseminate knowledge so that it can be used to improve the human condition. Effective communication thus plays an essential role in promoting technical advances. Simply put, communication is the ability to express oneself in a way that is understood readily and clearly. There will be no impact of scientific or engineering discoveries without effective written and oral communication.

In sections on writing well, we focus primarily on style—that is, rules of usage as well as principles of composition and form—and draw heavily from Strunk and White (1979), Berry (1971), and Brogan (1973). Indeed, many illustrative phrases and sentences were inspired by or modified from these works. We thus note here our indebtedness to these outstanding works and the examples therein. We encourage the reader to consult these excellent resources as well.

Although written communication, particularly the archival journal article, is most important to the widespread and long-term advancement of science and engineering, oral communication plays a vital role. From didactic lectures by an instructor to entertaining presentations for a lay audience, oral communication has the potential to capture the imagination and promote the advancement of science and its applications. Similar to theater, oral communication requires one to "tell a story" well, that is, to package information in a way that is assimilated quickly and retained. Technical advances in audiovisual capability can aid tremendously in stimulating and capturing an audience and thus should be integrated thoughtfully within the oral presentation.

It takes a lifetime to establish a good reputation, but only a moment to destroy it. Integrity in the workplace is just as important as understanding well the basic principles of science and engineering or the basic operation of a scientific instrument. Yet, even within the narrow context of technical communication, it is impossible to articulate a prescriptive set of rules or procedures for acting ethically. Despite the increasing prevalence of courses in research ethics, surveys suggest that most students learn the ethics of research and communication primarily "on the job", principally from their research mentor. Good training in the ethics of research thus begins with selecting a mentor who values such training and seeks to develop integrity through regular discussion and introspection. One goal herein is to stimulate this process of interaction around major issues most likely to face scientists and engineers in documenting and reporting their research.

The overall goal of this short book is not to be a standalone source on matters of style (which is left to professors of English or communication) or ethics (which is left for professors of philosophy or law). Rather, it is meant to motivate the reader to develop an effective, individual style of communicating and a personal commitment to integrity simply because it matters. Hence, this book is written based on personal experiences of the authors in research and training in the biomedical sciences and engineering, including the development and delivery of related graduate courses at Texas A&M University, Columbia University, and University of Virginia. Nevertheless, one of the best ways to learn to write well is to read extensively the works of good writers; one of the best ways to learn to speak well is to listen carefully to good speakers; one of the best ways to ensure integrity in the workplace is to learn from reputable professionals. The reader is thus strongly encouraged in this regard and, indeed, to keep a notebook wherein personal experiences and helpful observations can be recorded and reviewed periodically. Best wishes.

Acknowledgments

Second Edition

We once again thank Mr. Joel Claypool who invited and guided this update as a Second Edition, now with Springer Nature publishing. We also thank our many colleagues and former students who both affirmed this work and provided suggestions for improvements.

First Edition

We thank Jodi Eyhorn, from the Department of Communication of Texas A&M University, for expert assistance in correcting early portions of the manuscript. We also thank Joel Claypool, of Morgan & Claypool, for patience and support during the writing process.

Portions of this work began via a Special Opportunity Award from the Whitaker Foundation. Finally, J.D.H. thanks Carolyn S. and Tommie E. Lohman for their continued support of many different educational initiatives at Texas A&M University, including composition of portions of this work.

Contents

About the Authors

Jay D. Humphrey Ph.D., is John C. Malone Professor and former Chair of the Department of Biomedical Engineering at Yale University. He has authored a graduate textbook (*Cardiovascular Solid Mechanics*), co-authored an undergraduate textbook with a former student (*An Introduction to Biomechanics*), and published over 30 chapters in other books and encyclopedias and more than 375 archival journal papers. He served as founding Co-editor in Chief for the international journal *Biomechanics and Modeling in Mechanobiology* and as a reviewer for over 100 different technical journals and over 30 funding agencies in the United States and abroad. Before moving to Yale, he was Carolyn S. and Tommie E. Lohman Professor at Texas A&M University where he received a College of Engineering Teaching Award for classroom instruction, including for the course *Communication and Ethics in BME*. He has served as a research mentor for more than 175 postdoctoral fellows, graduate students, medical students, and undergraduates.

Jeffrey W. Holmes M.D., Ph.D., is Dean of the School of Engineering at the University of Alabama at Birmingham where he also serves as Professor of Biomedical Engineering, Medicine, and Surgery. He began his career at Columbia University, where he helped to build a new Department of Biomedical Engineering, developed and taught the course *Ethics for Biomedical Engineers*, and was a Faculty Associate at the Center for Bioethics. At the University of Virginia, he served as Graduate Program Director of Biomedical Engineering and founded the Center for Engineering in Medicine, working to improve communication and collaboration among engineers and clinicians. Over the course of his career, he has served as a member and Chair of an Institutional Animal Care and Use Committee (IACUC); a faculty mentor for more than a dozen graduate programs and training grants; and a research mentor for more than 100 postdoctoral fellows, graduate students, medical students, and undergraduates.

Motivation

1

The first part of this book is about communicating well, which is just as important to success in the workplace in science and engineering as it is in professions such as business, law, politics, and theology. Although there are useful guidelines for communicating well, there are no unique formulas. Individual differences can bring a freshness and vitality to a field; individual personalities can generate excitement and interest. Each person should thus develop a style that is most effective for them. The second part of this book addresses the need for professional responsibility, that is, integrity in the workplace. It has been correctly said that it takes a lifetime to establish a good reputation, but only a moment to destroy it. There is, therefore, a need for consistent and concerted attention to ethical conduct and appropriate documentation and communication of research findings. This, too, requires thoughtful, personal commitment—it is more than simply trying to follow the rules, for rules may change, it is doing the right things for the right reasons. Consequently, this book is different from most textbooks in science and engineering. We seek to cause one to stop, contemplate, and develop personal habits. In some sections, therefore, the style is more like a workbook. Indeed, as a point of departure, let us consider the following.

Exercise 1.1

List five of the most important inventions of all time.

   1.

   2.

   3.

   4.

   5.

© The Author(s), under exclusive license to Springer Nature Switzerland AG 2024 1
J. D. Humphrey and J. W. Holmes, *Style and Ethics of Communication in Science and Engineering*, Synthesis Lectures on Engineering, Science, and Technology,
https://doi.org/10.1007/978-3-031-39125-5_1

Exercise 1.2
List five of the most important scientists, mathematicians, or natural philosophers of all time.
   1.
   2.
   3.
   4.
   5.

If you are like many of our previous students, you listed among your selections of the most important inventions of all time the airplane, automobile, computer, electric motor, refrigeration, steam engine, telephone, and so forth. Each of these inventions is truly great, and so too many others, but did you consider the invention of the printing press with movable type? If not, you are not alone. Yet, take a few minutes and imagine how civilizations would have developed without books or scientific periodicals. Indeed, think about how the development of science, engineering, and medicine may have differed over the past five centuries had the printing press not been invented. For this reason, *Time-Life* magazine selected the printing press as the most important invention of the second millennium.

The rapid growth of printing with movable type reveals its overall importance. Invention of movable type is generally attributed to Johann Gutenberg (ca. 1397–1468), and the first book so printed is the famous Gutenberg Bible, which was completed in 1454/1455 at Mainz. By 1480, 111 towns throughout Europe boasted printing presses, and by 1500, this number grew to more than 238 (Boorstin, 1983, p. 270). In addition to the printing of the Bible, which had a significant influence on the development of science and Western culture (Dillenberger, 1961; van Doren, 1991), these presses allowed the printing and widespread distribution of classics by Aristotle, Cicero, Euclid, Plutarch, Ptolemy, and many others. It is thus thought that Gutenberg's invention played a singularly important role in the European Renaissance.

Recalling Exercise 1.2, many scientists, mathematicians, and philosophers deserve recognition as great. Among them, you may have listed Socrates (470–399 BC), Aristotle (384–322 BC), Archimedes (287–212 BC), Copernicus (1473–1543), Galileo (1564–1642), I. Newton (1642–1727), L. Euler (1707–1783), A. Lavoisier (1743–1794), C.F. Gauss (1777–1855), C. Darwin (1809–1882), J.C. Maxwell (1831–1879), J.W. Gibbs (1839–1903), M. Planck (1858–1947), M. Curie (1867–1934), A. Einstein (1877–1955), E. Noether (1882–1935), L. Pauling (1901–1994), or G. Hopper (1906–1992). How is it that we know about these great investigators? How is it that we know what they accomplished? Consider Sir Isaac Newton, for example, who is universally listed as one of the greatest natural philosophers of all time. Many know of Newton through textbooks on calculus or physics related to his law of gravitation, his laws of motion, or perhaps his experiment with a glass prism that revealed a spectrum of colors in sunlight. Fewer people know about Newton through in depth study, for example, by reading books such as *The Life of Isaac Newton* by Westfall (1993). Still fewer yet know of Newton because they have read his great works, his *Principia* of 1687 or his *Opticks* of 1704. Regardless of the particular path, we all know of Newton primarily through the written word, not oral tradition and certainly not first-hand interaction.

When reading about Newton, it is interesting to learn that he abhorred criticism of any kind and, in particular, interpersonal conflicts. Indeed, it appears that he was so concerned about criticism, especially from Robert Hooke (1635–1703), then secretary of the Royal Society of London, that for many years Newton had little interest in publishing his greatest ideas. Apparently, the *Principia* was published only because of the persistent encouragement and personal financial sacrifice by Edmund Halley (1656–1742). This is remarkable! Similarly, it seems that Newton withheld publication of his *Opticks* until just after the death of Hooke. What if Newton had died first? Can we imagine how the development of science may have differed had Newton not revealed his brilliant thoughts through these two books?

Likewise, it is interesting to contemplate the development of Western society, which has depended so strongly on Greek thought, had it not been for Plato (ca. 427–347 BC). It seems that Plato's mentor, the great Socrates, was content to lecture and discuss rather than to write or dictate. Although it is not clear how much of Plato's writings truly reflect Socrates, the importance of works such as *The Republic* is without question. The written word and its widespread distribution has impacted the world like few other things—it is fundamental to communication among scientists and engineers as well as the general public.

> Communication is defined as follows: To make known; impart. To transmit; have an interchange, as of ideas. To express oneself in such a way that one is readily and clearly understood.
>
> *American Heritage Dictionary*

Whether to have a long-lasting impact on human history or simply to contribute to success in the workplace, communicating well is a vitally important skill for the scientist and engineer. Not only must one communicate well with colleagues or a technical boss, there is often a need to communicate with diverse scientists, engineers, clinicians, venture capitalists, or the general public. For example, the National Institutes of Health (NIH) has promoted the importance of "team science" in biomedical research, which depends strongly on effective communication among individuals having diverse backgrounds, and many universities in the United States have promoted the importance of translational research, which in the biomedical realm requires interactions among clinicians, scientists, engineers, and those in business. Hence, we cannot overemphasize the importance of effective written and oral communication in research and development.

Because our focus is on technical communication, note that 6 March 1665 marks a beginning of the scientific periodical, based on the proceedings of the Royal Society of London entitled *Philosophical Transactions*. In the preface to the first issue, Henry Oldenburg (ca. 1617–1677) wrote (see Boorstin, 1983, p. 393):

> Whereas there is nothing more necessary for promoting the improvement of philosophical Matters, than the communicating of such, … It is therefore thought fit to employ the *Press*, as the most proper way to gratified those, whose engagement in such Studies, and delight in the advancement of Learning and profitable Discoveries, doth entitle them to the knowledge of what this Kingdom, or other parts of the World, so, from time to time afford. . . .All for the Glory of God, the Honor and Advantage of these Kingdoms, and the Universal Good of Mankind.

A careful editor would likely revise this Preface in the interest of conciseness and clarity, yet the message would remain and so too the motivation to improve our communication skills. Indeed, from its beginning, the Royal Society of London sought clear communication in both written proceedings and oral presentations, not an "Artifice of Words." We are well advised to pursue the same today.

In conclusion, recall from the Preface that a good way to learn to write well is to read widely. The student having an interest in engineering, mathematics, medicine, philosophy, or science can learn much from the many books on the historical development of these fields. See Shamos (1959), for example, who provides brief background information on great physicists from Galileo to Einstein and includes excerpts from their original publications. Lightman (2005) provides a similar resource for scientists of the twentieth century, and Clendening (1960) reprints portions of the early great publications in medicine. Other books of interest include Bell (1986), Mason (1962), Motz and Weaver (1989), Tarnas (1991), and van Doren (1991). It is interesting to conclude, consistent with aforementioned comments by Boorstin (1983), that Shamos (1959) observed: "The exchange of knowledge, facilitated by the publication of scientific journals, became—and remains—one of the most significant factors in the growth of physical science."

Exercise 1.3
Write and submit a three-page (double-spaced, one-inch margins, 12-point font) essay on the role of printed books and scientific periodicals during the Age of Enlightenment.

Exercise 1.4

Write and submit a three-page (double-spaced, one-inch margins, 12-point font) essay on the development of the Royal Society of London and its role in the advancement of science.

Exercise 1.5

Write and submit a three-page (double-spaced, one-inch margins, 12-point font) report on the origins of the university and how it differed from the scientific academies of the 17th and 18th centuries.

Exercise 1.6

Write and submit a three-page (double-spaced, one-inch margins, 12-point font) biography of a famous philosopher, mathematician, scientist, or engineer whom you would like to learn more about. If possible, include in your literature research at least one primary source.

References

Bell, E. T. (1986). *Men of Mathematics*. Simon & Schuster.

Boorstin, D. J. (1983). *The Discoverers*. Vintage Books.

Clendening, L. (1960). *Source Book of Medical History*. Dover Publications.

Dillenberger, J. (1961). *Protestant Thought and Natural Science*. University of Notre Dame Press.

Lightman, A. (2005). *The Discoveries: Great Breakthroughs in 20th Century Science*. Vintage Books.

Mason, S. F. (1962). *A History of the Sciences*. Collier Books.

Motz, L., & Weaver, J. H. (1989). *The Story of Physics*. Avon Books.

Shamos, M. H. (1959). *Great Experiments in Physics: Firsthand Accounts from Galileo to Einstein*. Dover Publications.

Tarnas, R. (1991). *The Passion of the Western Mind*. Ballantine Books.

Van Doren, C. (1991). *A History of Knowledge*. Ballantine Books.

Westfall, R. S. (1993). *The Life of Isaac Newton*. Cambridge University Press.

Writing Well

2

> *Vigorous writing is concise. A sentence should contain no unnecessary words, a paragraph no unnecessary sentences, for the same reason that a drawing should have no unnecessary lines and a machine no unnecessary parts. This requires not that the writer make all his sentences short, or that he avoid all detail and treat his subject only in outline, but that every word tell.*
>
> Strunk and White (1979, p. xiv)

2.1 Overall Approach

Two of the most difficult aspects of writing are getting started and finishing well. In other words, once we begin in earnest, it is usually easy to continue our line of thinking and to produce a first draft. Revising and polishing the first draft often takes longer than the initial writing, yet this is time well spent. Consider, therefore, some general guidelines for writing well, including a simple five-step recipe for completing a technical document:

1. Outline in detail.
2. Write freely.
3. Edit critically.
4. Read out loud.
5. Have a colleague proofread.

© The Author(s), under exclusive license to Springer Nature Switzerland AG 2024
J. D. Humphrey and J. W. Holmes, *Style and Ethics of Communication in Science and Engineering*, Synthesis Lectures on Engineering, Science, and Technology,
https://doi.org/10.1007/978-3-031-39125-5_2

Although these steps may seem obvious, even trivial, they serve as important reminders and aid greatly in the composition of each new work no matter the level of one's experience.

2.1.1 Outline

Most writers agree that it is useful to begin with a detailed outline. Such an outline should contain the major headings that guide the flow of the work, but perhaps more importantly, it should also contain subheadings with bullets that highlight and order the major points within each section.[1] Each document is different, thus we should not feel compelled to force our presentation to fit within the confines of a particular outline. Nonetheless, most technical works adhere to the following basic outlines:

An illustrative technical proposal

- Project Summary
- Specific Aims
- Significance
- Innovation
- Research Plan
- References

A representative technical paper or report

- Abstract
- Introduction
- Methods
- Results
- Discussion
- References

The traditional M.S. thesis or Ph.D. dissertation[2]

- Abstract
- Introduction

---

[1] Some encourage full sentences rather than bullets to document the main ideas in each subsection or paragraph, sentences that may later be used directly in the document. This, however, is a matter of personal style.

[2] Many European dissertations are very different. They consist of a sequence of individual chapters, each similar to a technical paper, all of which are tied together via short introductory and concluding chapters.

- Background
- Methods
- Results
- Discussion
- Conclusions and Recommendations
- References
- Appendices

Because of their importance, we discuss these different forms of technical communication in detail in Chaps. 3 and 4. Here, we simply emphasize that a basic outline is the first step toward successful writing; it organizes the flow of the presentation and reminds us to address particularly important issues. Additional bulleting within each subsection further directs the writing. Indeed, one may easily use the final outline as a beginning document.

In summary, as in any activity, we increase our chances of reaching our destination when we have a map or detailed instructions that show the way. Note, therefore, that an outline will tend to be most useful and focused when it is constructed against the background of two questions (Gibaldi, 1995):

What is the overall goal that you wish to achieve with the document?
Who is the intended audience?

Toward this end, it is useful to critique the final outline with regard to both consistency and conciseness. For example, do points made in the Introduction set up well the key points made in the Discussion? Moreover, we tend to allow space in our outline for all of the information that we have collected during our research; we should, however, delete information that unnecessarily duplicates or distracts, that is, we should avoid material that is irrelevant or unnecessary. Once done, it is then time to begin the actual writing.

2.1.2 Write Freely

One of the greatest impediments to writing efficiently and effectively is untimely self-criticism. How many of us have labored over that first sentence or first paragraph, rewriting and editing to the point of fatigue or frustration? Such editing is essential, but it is productive only if addressed at the right time and in the right way. By writing freely, we mean the unencumbered recording of a logical thought process. Indeed, it is often useful to disable the spell- and grammar-checking capabilities of word processors during the initial writing, for they contribute to the distractions of worrying about the initial spelling of words, ordering of phrases, and even punctuation. These and similar issues are addressed easily once we complete the initial draft. We likewise should not initially worry about emphasizing active voice, ensuring sufficient variety in our word choice, focusing on conciseness, and so forth. Rather, at this early stage of composition, the most important goal is to ensure that the major ideas are recorded and organized roughly in the right order.

2.1.3 Edit Critically

Once the first draft is finished, it is usually best (if time permits) to put it aside for at least a few days before beginning to edit critically. The reason for this is that we often see what should be there rather than what is there when we proofread our documents. Most of the remainder of this chapter addresses specific aspects of editing critically, which typically includes adding, deleting, and rearranging text.

The fundamental components of any technical document are sentences and paragraphs. A sentence is a grammatical unit typically consisting of a subject and a predicate (which tells something about the subject). A simple example is—I am. R. Descartes (1596–1650) expanded this to read—I think, therefore I am. Clearly, important and memorable sentences need not be complex. A paragraph is a grammatical unit typically consisting of multiple sentences that together express a complete thought. Many suggest that the lead sentence of each paragraph should introduce the main idea of that paragraph and the final sentence of each paragraph should summarize the main thought. This simple guideline helps to minimize unnecessary generality, that is, it helps to keep the writer focused.

A stepwise approach to editing critically exploits these two fundamental units of composition. For example, many suggest that the first step of editing should consist of reading the first and last paragraphs of the document to ensure a consistent Introduction and Conclusion. The second step should consist of reading the first sentence of each successive paragraph to ensure that the work flows logically. Indeed, some go further to suggest that one should be able to glean the salient points of a document by reading only the first sentence of each successive paragraph. Although we do not suggest such a dogmatic approach, casual guidance can come from such an exercise. The third step of critical editing is a careful evaluation sentence by sentence. In other words, while reading each sentence within context, we should ask if it is necessary, if it is consistent in tense, and if it as concise and clear as possible. This brings us to the fourth and last step of critical editing, an evaluation word by word. We should ask, for example, if we have avoided the use of jargon as well as redundant or unnecessary words and if the intended meaning of each word reflects its definition. Word choice is critical. From a pragmatic perspective, we can simultaneously evaluate sentence by sentence and word by word.

As noted previously, the importance of critical editing cannot be overemphasized, hence we return to this issue in detail in Sects. 2.2–2.8. Here, however, let us finish our discussion of an overall approach to writing well. After we have outlined our work, written freely, and edited carefully, our next step should be to read the document out loud.

2.1.4   Read Out Loud

Although this step may seem trivial or perhaps uncomfortable, it is amazing how sensitive the ear is to effective writing—different tenses, illogical sequencing, unintentional rhymes, the overuse of certain words, and so forth. We strongly recommend, therefore, that one read the document out loud before going to the final step, asking a colleague to provide constructive criticism.

2.1.5   Have a Colleague Proofread

Technical advances in science and engineering have been spectacular, and continued promises of important discoveries make these professions intellectually attractive and socially critical. Because of the trend toward multidisciplinary teams, one of the most enjoyable aspects of these professions is the opportunity to work with diverse colleagues from many allied fields. Consequently, there is not only a need to write concisely, but also to write clearly. Although it is common to have colleagues from these allied fields coauthor many of our works, it is essential to have others proofread our work. That is, even though we may know best what needs to be said, the definition of an effective paper, proposal, report, or book is one that is understood and valued by others—this is the goal of effective communication. Classmates and colleagues tend to be busy, thus we sometimes hesitate to ask them to proofread our work. Yet, they too would appreciate having someone provide feedback on their work and consequently will many times agree to do so for you. Consider establishing reciprocal agreements whereby you exchange documents to be proofread. Such reciprocity will not only help the author by providing specific feedback, chances are it will help the reader both directly and indirectly. One not only learns by reading, oftentimes going through a document carefully and looking for ways to improve its clarity and conciseness teaches us much more. This is similar to the adage: "While we teach, we learn" (Seneca, 4 BC–65 AD).

If you ask someone to proofread your work, make sure to tell them that you want them to be brutally honest rather than overly concerned about being critical. Moreover, once you receive the feedback, be careful to avoid the two most common responses: either to ignore the comments because you know that you were correct in the first place or to incorporate the suggested changes without questioning. These two responses are equally inappropriate. If a colleague questions the way something is stated, particularly if based on deep knowledge of the technical area, this at least suggests that the text could be written more clearly. In other words, if they do not understand what you are trying to say, chances are that others will likewise not understand. Consider revising the text along the lines they suggest or in another way; the important thing is to give the suggested revision careful consideration. Conversely, incorporating suggested changes without questioning is dangerous. The primary goal is to communicate most effectively that which you are trying to say. If your colleague's suggestion does that, great; if not, work to improve clarity and conciseness and perhaps have your colleague read it again. Often, it is helpful to ask them what was confusing or what they thought you meant to say. Sometimes an explanation reveals how best to say it in the written word.

Finally, remember that it is good to keep marked manuscripts to evaluate them for possible consistent errors or patterns. In this way, we can become proactive in avoiding problems, whether it is an overuse of passive voice or an inappropriate use of modifiers. Being conscious of potential errors is the first step to avoiding them.

> **Exercise 2.1**
> Recover from your files at least three documents that you have written and that were evaluated/graded on writing style. Compile a list of errors that occurred repeatedly and write brief examples of how to correct these problems.

> **Exercise 2.2**
> Read a journal paper in your area of expertise and record at least 10 sentences that could be improved for conciseness or clarity. Next, reread the paper out loud and record another five sentences that you could improve; note the types of concerns that are identified more easily when heard. Finally, suggest possible improvements for each of the 15 sentences.

2.2   Removing Redundancies and Unnecessary Words

Now that we have a feel for an overall approach to writing well, let us begin to address specific aspects of critical editing. Recall that effective technical writing is first and foremost clear and concise, which for obvious reasons is better written "Effective technical writing is clear and concise." One way to ensure such characteristics in our writing is to *remove redundancies and unnecessary words*, sentence by sentence. Let us consider a few specific examples below (note: the original version is on the left and the corrected version is on the right, hence it is best to cover the right side first and consider how you might improve each example before looking at the suggested change):

The cells were cultured for a period of three weeks.	The cells were cultured for three weeks.
The temperature of the chamber remained between 35 and 39 °C.	The chamber remained between 35 and 39 °C.

The associated mechanisms are not known at this time.	The associated mechanisms are not known. (or,… remain unknown.).
The experiments were performed over a period of 10 h.	The experiments were performed for 10 h.
The new transducer is much smaller in size, which simplifies the design.	The new transducer is smaller, which simplifies the design.
The temperature increased at a rate of 3°/min.	The temperature increased at 3°/min.
The signal is lost below a threshold level of 10 Hz.	The signal is lost below a threshold of 10 Hz.
This thesis reports work done during the period from January 1998 to December 2000.	This thesis reports work accomplished from January 1998 to December 2000.
The algorithm searches outward from the center location.	The algorithm searches outward from the center.
The A/D converter allows a maximum of eight input signals.	The A/D converter allows eight input signals.
The range of the output signal was from a minimum of 2 to a maximum of 5 V.	The output signal ranged from 2 to 5 V.
The results of our experiments support the established theory.	Our experiments support the established theory.
Turn the potentiometer in the clockwise direction to increase the gain.	Turn the potentiometer clockwise to increase the gain.
Use the lenses that are convex in shape.	Use lenses that are convex.
The problem should first be formulated and then solved.	The problem should be formulated, then solved.
The amount of noise will be excessive if the signal is not filtered.	The noise will be excessive if the signal is not filtered.
There is no known analytical solution to this equation at this time.	No analytical solution exists for this equation.
The biopsy should be redesigned in the future to minimize the amount of tissue needed.	The biopsy should be redesigned to minimize the tissue needed.
The reason for this difference can be attributed to….	This difference can be attributed to….
Remember to remove the specimen during the calibration procedure.	Remember to remove the specimen during calibration (or, when calibrating).
There is a growing body of evidence that the hypothesis is indeed true.	There is growing evidence that the hypothesis is true.
After one test, there should be a sufficient quantity of culture media for a second test.	After one test, there should be sufficient culture media for a second test.

In the first example, the use of the word weeks implies a period or duration, which is therefore not needed. In the second example, the use of the unit °C specifies that the numerical value refers to a temperature, which thereby becomes redundant. Similarly, in the third example, we see that if something is not known, it is implied that it is not known at this time, which is thereby unnecessary. In hindsight, the other examples are likewise clear. Indeed, because these specific examples highlight a redundancy or unnecessary words, they may seem so obvious that we would be surprised if we ever wrote such sentences. Upon close examination of our previous works, however, we often find similar or even more flagrant examples. It is for this reason that we must be conditioned to look for redundancies and unnecessary words, which is often best learned via examples; see Brogan (1973) for additional examples.

Exercise 2.3
Find a technical research paper that you have written and scan it specifically for examples of redundancies or unnecessary words. Record five examples below with possible revisions.

If we read published technical papers for style, we quickly realize that we could render many commonly used phrases more concise or even omit them without a loss of clarity. For example, how many times have we read the phrase "The purpose of this paper is to present…," which we could write more concisely as "This paper presents…" Consider the following phrases (left side) that occur frequently but can often be edited or omitted (right side) as follows (cf. Brogan, 1973; Valiela, 2001):

Is used to develop	Develops
Is dependent on	Depends on

We propose to use the combination of	We propose to combine
Results in the simplification of	Simplifies
It is interesting to note that	Note that (or, omit)
Due to the fact that	Because
In order to	To
In spite of the fact that	Despite
As a result of	Omit
Appears to be	Seems
Experienced a peak at	Peaked at
In the event that	If
Was found to be	Was
A number of	Many (or, various)
May be a mechanism responsible for	May cause
It is well-known that	Omit
For a long period of time	For a long period
Is described in detail in	Is detailed in
In the absence of	Without
It is not uncommon that	It is common that
The finding is not inconsistent with	The finding is consistent with

Note, in particular, that the last two entries in this table emphasize that double negatives should be avoided. Moreover, see Appendix 2 in Day and Gastel (2006) for an expanded list of words and expressions to avoid.

In addition to removing redundancies and unnecessary words, there are many opportunities to *introduce conciseness* via word choice and sentence structure. For example, consider the following examples:

Because the structure is assumed to remain circular,...	Assuming the structure remains circular,...
This will enable us to develop a better understanding of...	This will enable us to understand better...
This finding is the opposite of that reported by...	This finding is opposite that reported by...
The model is capable of describing...	The model can describe...
Table 1 provides a list of all findings...	Table 1 lists all findings...
The next section is a brief description of the experimental methods	The next section briefly describes the experimental methods
The faculty advisor was the supervisor of both the undergraduate and the graduate students	The faculty advisor supervised both the undergraduate and the graduate students

The temperature readings will be dependent upon the contact stress…	The temperature readings will depend on the contact stress…
Our laboratory technician also serves as the budget manager	Our laboratory technician also manages the budgets
The following example is an illustration of the basic concepts of…	The following example illustrates the basic concepts of…

Before continuing, note that some of the suggested changes in the right-hand column assume a particular style not accepted by all technical writers. Some suggest that a table cannot list, a figure cannot show, a model cannot describe, a paper cannot present, and so forth. That is, some argue that only people can list, show, describe, or present; results are simply listed in a table, shown in a figure, described by a model, or presented in a paper. Because this is a matter of style, one must decide what approach to take, then be consistent.

As we shall see in Chaps. 3 and 4, removing redundancies and unnecessary words not only results in writing that is crisp, clear, and concise, it enables us to meet stringent limitations on words or pages in published works or proposals. Consider two instructive exercises:

Exercise 2.4
Write a three-page (double-spaced, one-inch margins, 12-point font) biography of a contemporary scientist. Two days after finishing your essay, edit it further to reduce it to a single page without losing significant content. You will be surprised how easy and yet how powerful it is to write for conciseness. Finally, note that a one-page white paper is often all that is used to render important decisions in many professions; it is thus important to write effective short reports.

Exercise 2.5
Use the "Word Count" tool in your word processor to determine the number of words in a short document (e.g., abstract) that you recently composed. Once done, reduce the length of the document by 50% without compromising the content.

2.3   Active Voice, First Person, and Different Tenses

2.3.1   Voice

In *active voice*, the subject of the sentence performs the action indicated by the verb. Conversely, in *passive voice*, the subject of the sentence receives the action of the verb. The simple example below distinguishes between passive and active voice:

Passive voice: The data were analyzed by us using an ANOVA.[3]
Active voice: We analyzed the data using an ANOVA.

Although passive voice is acceptable, indeed sometimes more appropriate, most technical writers prefer active voice for it engenders conciseness and directness. In the example given, we see that seven words suffice rather than nine—this reduction represents a savings of approximately 20%. Given a 10-page paper, a 20% reduction in the number of words would yield an eight-page paper or else would provide an extra two pages to include more information; such savings can be significant. Moreover, comparing the two sentences in this example reveals the increased directness of the active voice, which promotes clarity and conciseness.

Albeit preferred, active voice is less common than passive voice in scientific writing. A simple change in the preceding example illustrates one reason for this:

Passive voice: The data were analyzed using an ANOVA.
Active voice: We analyzed the data using an ANOVA.

In this case, each sentence contains seven words, thus the active voice does not increase conciseness. Moreover, the context can imply the "we" in the case of the passive voice, thus there need not be a difference in clarity (e.g., who did what). In many cases, authors prefer not to write in the first or the third person and revert to passive voice. The issue of person is addressed in the next section (or should we say, we address the issue of person in the next section). Here, however, consider examples of passive voice (left) and easy ways of changing them to active voice (right). First, changes primarily in verb form can be effective:

The specimen is connected to the device through a custom cannula	The specimen connects to the device through a custom cannula
The output signal is fed into a signal conditioner	The output signal feeds into a signal conditioner
In the next section, the underlying theory is given	The next section gives the underlying theory

---

[3] ANOVA is a common acronym for analysis of variance in statistics.

In our current research, attention is directed to finding the mechanism	Our current research focuses on finding the mechanism
The theory is dependent on five basic postulates	The theory depends on five basic postulates
X was used to create a surface-confined computational mesh	X created a surface-confined computational mesh
Increasing evidence has implicated the importance of…	Increasing evidence implicates the importance of…
Three different sectioning planes were used to form…	Three different sectioning planes formed…
Experimental noise is increased when unshielded cables are used	Experimental noise increases with the use of unshielded cables
A reader's attention is increased by the liberal use of figures and schematic drawings	A reader's attention increases with the liberal use of figures and schematic drawings

Second, changing the subject, which often necessitates changing the order of the words in the sentence, is often equally effective:

The specimen is connected to the device through a custom cannula	A custom cannula connects the specimen to the device
The results of the study are listed in Table 1	Table 1 lists the results of the study
The control is simplified by using commercial software	Commercial software simplifies the control
An improved result is obtained by refining the computational grid	A refined computational grid improves the result
The proper use of the equipment is described in Chap. 2 of the manual	Chapter 2 of the manual describes the proper use of the equipment
Ensure that all specimens are tested under the same conditions	Use the same conditions to test all specimens
The temperature is measured by a thermocouple	A thermocouple measures the temperature
These empirical findings are used as inputs into the theoretical model	The theoretical model uses the empirical findings as inputs
A detailed derivation of this equation is given in the appendix	The appendix details the derivation of this equation
The culture system is optimized by maintaining body temperature	Maintaining body temperature optimizes the culture system

A common excuse for (over)using passive voice is that it is natural because we often discuss past events, as, for example, "it was reported that" or "it was found that." Tense need not be the deciding factor, however, as revealed by the following simple example:

The pressure was measured by a mercury manometer	A mercury manometer measured the pressure
The pressure is measured by a mercury manometer	A mercury manometer measures the pressure

Finally, as noted previously, writing in the first or third person often allows us to avoid passive voice. Whereas the next section discusses the issue of person, as appropriate in scientific writing, consider the following, which revisit previous examples with alternate changes:

The specimen is connected to the device through a custom cannula	We connected the specimen to the device using a custom cannula
The results of the study are listed in Table 1	We list the results of the study in Table 1
These empirical findings are used as inputs into the theoretical model	We used the empirical findings as inputs into our theoretical model
The temperature is measured by a thermocouple	We measured the temperature using a thermocouple
A detailed derivation of this equation is given in the Appendix	I derive this equation in detail in the Appendix
Experiments were performed in triplicate for each set of…	We performed three experiments for each set of…
The culture system is optimized by maintaining body temperature	We optimized the culture system by maintaining body temperature

In summary, we do not need to avoid passive voice at all costs; indeed, passive voice is preferred in many cases. We also need not invoke first person to avoid passive voice. Nevertheless, our general guideline is to prefer active voice when editing critically.

Exercise 2.6
Select a journal paper in your field that interests you and scan it specifically for examples of passive voice. Record five examples below with possible revisions.

2.3.2 Person

Students of the history of science know that scientific writing used to be much more personal. As a simple example, consider the following excerpt from one of the works of W. Harvey (1578–1657) on the motion of the heart (Clendening, 1960, p. 159):

> Besides the motions already spoken of, we have still to consider those that appertain to the auricles. Casper Bauhin and John Riolan, most learned men and skillful anatomists, inform us from their observations, that if we carefully watch the movements of the heart in the vivisection of an animal, we shall perceive four motions distinct in time and in place, two of which are proper to the auricles, two to the ventricles. With all deference to authority I say, that there are four motions distinct in point of place, but not of time. ...

If written today, we may well have read (with little other editing):

> Besides the motions already noted, there is a need to consider those concerning the auricles. Bauhin (16xx) and Riolan (16xx) report that careful monitoring of the heart in an open-chest animal reveals four motions distinct in time and place, two of the auricles and two of the ventricles. Nevertheless, it is suggested here that these four motions are distinct in place but not time. ...

Why has scientific writing become so impersonal today? Certainly, there has been an appropriate move away from the verbose, from patronizing prose, and from self-aggrandizement. Nevertheless, science and engineering are personal—they are advanced by people, usually for the good of people—and it is not only acceptable to write in first person, in many cases it is more honest, direct, and effective. For example, in Chap. 4 on writing research proposals, we will see that an important part of the NIH-R01 grant is a section called Innovation. Imagine that you review such a section and read, "It has recently been shown that… (12)." Noting that (12) denotes reference number 12 in the list of references, the reviewers would not know if it was the applicant or another investigator who showed this important finding unless they looked at the reference list. Conversely, there would be no ambiguity if the applicant wrote "We recently showed that… (12)." In the case of a research proposal, clearly demonstrating one's previous work may increase tremendously the chances of funding, thus employing first person may be both effective and advantageous.

As a reminder that first person can yield effective and memorable prose, recall the following sentences from the seminal paper by James Watson (1928-pres) and Francis Crick (1916–2004) on the structure of DNA:

> We wish to suggest a structure for the salt of deoxyribose nucleic acid (D.N.A.). … It has not escaped our notice that the specific pairing we have postulated immediately suggests a possible copying mechanism for the genetic material.

There are three persons in typical grammatical structure: *first person* refers to the person or persons who are speaking or writing; *second person* refers to the person or persons spoken or written to; and *third person* refers to person(s) spoken or written about. For example, consider the traditional pronouns, singular and plural, in the three persons and three common cases (Vivian & Jackson, 1961):

Singular	First person	Second person	Third person
Nominative	*I*	*You*	*He, She, It*
Possessive	*My, Mine*	*Your, Yours*	*His, Hers, Its*
Objective	*Me*	*You*	*Him, Her, It*

Plural	First person	Second person	Third person
Nominative	*We*	*You*	*They*
Possessive	*Our, Ours*	*Your, Yours*	*Their, Theirs*
Objective	*Us*	*You*	*Them*

Whereas the words *he* or *him* were used generically in the past to denote males or females, modern writers should be more sensitive to issues of gender. Thus, there has been a move to use neutral pronouns. For example, the famous imperative from Star Trek fame, "To boldly go where no man has gone before," could be written as "To boldly go where no one has gone before." It is possible to write *he/she, him/her, his/hers*, but neutral constructions *they, them, their* can be used as well. Finally, numerous terms such as *department chairman* or *layman* should be written as *department chair* or *layperson* to avoid this issue.

Albeit largely a matter of style, we suggest that it is acceptable and many times preferable to use a personal style in scientific writing. For example, it is acceptable to write: "Although they were the first to exploit their novel empirical observations by identifying quantitative correlations, we were the first to develop a theoretical basis to explain the observations."

As food for thought, consider the following simple examples as you decide on a particular style:

The authors recommend, therefore, that…	We recommend, therefore, that
Hence, it is suggested that…	Hence, I suggest that…
It will be seen that…	You will see that… or We will see that…
Based on these results, it was decided that…	Based on these results, we decided that…
It has been shown previously that…	We previously showed that…

One important warning, however, is that when "I" is used, be careful not to give the impression that it serves an egotistical end.

2.3.3 Tense

Tense is a property of time; it signifies when events occur or when conditions exist (Vivian & Jackson, 1961). There are six tenses: *past, present, future, past perfect, present perfect*, and *future perfect*. The perfect tenses typically involve the use of the words "have" or "had." Consider the following simple examples:

Past: I completed the experiment.
Present: I am completing the experiment.
Future: I will complete the experiment.
Past perfect: I had completed the experiment.
Present perfect: I have completed the experiment.
Future perfect: I will have completed the experiment.

Two of the key questions in scientific writing are, "What tense should I use when reviewing what others reported previously?" and "What tense should I use when reporting what I did?" Although it is of little comfort, the answer to these questions is that there is no set answer.

Some authors suggest, however, that if concepts or findings reported in a previous peer-reviewed work remain true, one should refer to them in the present tense. As a simple example, consider Newton's second law of motion, which was put forth in the seventeenth century. One could write "As Newton showed in the *Principia*, force equals mass times acceleration." Alternatively, one could write "As Newton showed in the *Principia*, force equaled mass times acceleration." Most agree that if it is still believed that force equals mass times acceleration, then present tense should be used. A more modern example could be, "Smith et al. (2021) show that..." versus "Smith et al. (2021) showed that..." Again, the choice is largely a matter of personal style; the most important thing is to be consistent within a given paper.

Most authors yet agree that we should use past tense when reporting our own new findings, for they have not yet been verified or accepted widely. Hence, when writing the results section of a paper, it is appropriate to use "we measured" and "we found" or similar constructs.

2.4 Infinitives and Modifiers

2.4.1 Infinitive

An infinitive is a verb form, a characteristic sign of which is the word *to*, for example, "to measure," "to quantify," or "to report" (Vivian & Jackson, 1961). A split infinitive occurs when a word or phrase separates the "to" and its complement. A famous split infinitive comes from the aforementioned quote from Star Trek: "To boldly go where no man has gone before," which we could rewrite as "To go boldly where no one has gone before." The issue is how we wish to go, boldly or fearfully. Although it is best not to split infinitives, grammarians are now less dogmatic with regard to this rule. Indeed, a purposefully split infinitive may be preferred in some cases. For example, consider the phrase "to promote exercise vigorously" (Iverson et al., 1998). There could be confusion by some as to whether vigorously relates to promote or exercise, hence writing "to vigorously promote exercise" could be clearer, unless of course the intent was "to promote vigorous exercise." Strunk and White (1979) also note that the sentence "I cannot bring myself to really like the fellow" is clear, concise, and relaxed. Nevertheless, the general rule should be: *Do not split infinitives unless the sentence is less awkward when doing so.*

Let us consider a few examples of split infinitives and how to correct them easily.

The goal of this project is to better understand...	The goal of this project is to understand better...
We plan to quickly initiate the funded study	We plan to initiate the funded study quickly
It is difficult to separately control X and Y...	It is difficult to control X and Y separately
..., they failed to correctly diagnose	..., they failed to diagnose correctly

It is bad practice in the laboratory to arbitrarily stop an experiment	It is bad practice in the laboratory to stop an experiment arbitrarily
To effectively study the source of the error,…	To study the source of the error effectively,…
The sponsor requested us to, with all possible haste, complete the final report	The sponsor requested us to complete the final report with all possible haste

The last example in this table is a particularly flagrant abuse of the infinitive.

Other examples of split infinitives occur when a single "to" serves multiple infinitives. Whereas it is generally acceptable to write, "There is a need to assemble and test the device," rather than "There is a need to assemble and to test the device," it is also better to write "There is a need to assemble the device according the sponsor's specification, then to test it…" rather than "There is a need to assemble the device according the sponsor's specification, then test it…"

Finally, note that infinitives can occur in active or passive voice and in past or present tense. In these cases, the infinitives may take different forms, such as:

Present active: to tell
Present passive: to be told
Past active: to have told
Past passive: to have been told

Hence, that a word or phrase appears between the "to" and its complement need not signal that an infinitive has been split.

2.4.2 Modifiers

Another mistake common in technical writing is the *use of nouns as modifiers*. A modifier is a word, phrase, or clause that renders another word or group of words more specific; two common kinds of modifiers are adjectives and adverbs. In contrast, a noun is a person, place, or thing. Perhaps it has been in the spirit of trying to write concisely that nouns have been frequently misused as modifiers. In a syndicated column, J.J. Kilpatrick noted a few examples from the *New York Times*: "their court victory," which is better written "their victory in court," and "close-knit classical music world," which is better written "close-knit world of classical music." Common examples in the technical literature include "material science" rather than "the science of materials" and "fluid mechanics" rather than "the mechanics of fluids." Yet, such constructions need not be considered problematic, which reminds us that certain cases are acceptable. More flagrant examples of noun modifiers exist in many scientific papers and should be minimized.

Tabulated below are a few examples found in recently published works (which we do not cite so as not to criticize particular authors, for many, including us, are equally guilty):

The primary extracellular matrix components include…	Primary components of the extracellular matrix include…
When tissue temperature reached…	When the temperature of the tissue reached…
Force and length data were used to compute stresses	Stresses were computed from data on forces and lengths
An increased wall stiffness of the aorta…	An increased stiffness of the wall of the aorta…
Minimum residual microfibrillar function…	Minimum residual function of the microfibrils…
Could detect molecular level changes…	Could detect changes at the molecular level…
Will use gene expression measurements to…	Will use measurements of gene expression to…
Changes in cell structure and function reveal…	Changes in structure and function of cells reveal…
The resulting surface stress appears…	The resulting stress at the surface appears…
The ability of the cells to move into the wound area…	The ability of the cells to move into the area of the wound…
To undergo changes in contractile protein expression…	To undergo changes in expression of contractile proteins…
Organ development becomes highly sensitive to…	Development of the organ becomes highly sensitive to…
Of the neonatal fibroblast…	Of the fibroblast in neonates to…

In contrast to previous tables of examples on redundancies, the "corrected" right-hand side here often resulted in a longer sentence or phrase. Again, it may have been in the interest of conciseness that nouns have come to be misused (left side). Nevertheless, one is well advised to use nouns properly.

Next, consider a few simple suggestions to promote the proper use of appropriate modifiers (adjectives and adverbs). Whereas adjectives modify nouns, adverbs can modify a verb, an adjective, or another adverb. Adverbs may come before, after, or between the words that they modify. When possible, a sequence of modifiers should be listed according to length or logical order. For example, Berry (1971) suggests that "tired, bored, and exhausted" is written better as "bored, tired, and exhausted" because it is likely that one becomes bored before tired and tired before exhausted. He likewise suggests that the modifiers "dry, withered, and flaky" should be ordered in the sequence in which they occur: "withered, dry, and flaky."

Finally, note that "a," "an," and "the" are articles. The definite article "the" refers to something or someone in particular. Hence, when we read "A significant finding was…" versus "The significant finding was…," we see that the former refers to one of many significant findings, whereas the latter refers to one finding that was significant. This simple distinction must be respected.

Although many modifiers are effective in different forms of writing, their overuse in scientific writing may suggest that one's results are not quantitative, that they need embellishing. For example, one may write that their numerical method is "very robust," but if it is robust, that is all that needs to be said. Similarly, instead of saying that data "are very noisy," we need only say that they "are noisy," then provide specific measures such as a signal-to-noise ratio to quantify the degree of the noise. A particularly flagrant use of "very" includes "this finding is very unique." Unique implies one of a kind, hence one should not modify it with very, mostly, nearly, or similar words. In summary, many modifiers like "very" often add very little (as in this case). Similarly, "quite" is quite unnecessary in most cases, including this one. A general rule, therefore, is: *Do not use modifiers unless the meaning is clarified by doing so.*

Exercise 2.7
Review a technical paper that you wrote previously and eschew all unnecessary or inappropriate modifiers. If you are somewhat puzzled why you used words such as "somewhat," take comfort that you are not alone. As examples, record five illustrative sentences below and possible corrections.

2.5   Additional Issues of Word Choice

The best advice related to word choice is to consult a dictionary frequently. With regard to the five-step recipe for composition given in Sect. 2.1, however, we should remember that this should be done during the phase, "edit critically." Indeed, if you are struggling for just the right word while "writing freely," it is often best to put an "xx" in the text so that you are reminded to search for an appropriate word later and not interrupt the flow of your thoughts and composition. Here, however, we briefly identify and discuss some words that are often used interchangeably but should not be so used. Consider, for example:

Alternative/alternation: An *alternative* is a choice between two mutually exclusive possibilities. An *alternation* is a successive change from one thing to another and back again.

Amount/number: *Amount* refers to a quantity that is not countable, whereas *number* is used when it is possible to count. It is thus correct to say "The amount of information available was not sufficient for…" or "The data suggest a number of conclusions."

Because/since: Strictly speaking, *because* refers to a cause–effect relationship and *since* refers to a past event. It is appropriate to write "Because the results suggested…" and "Since the last conference…"

Between/among: In general, use the word *between* when considering two things and use the words *among* or *amongst* when dealing with more than two things.

Can/may: The word *can* has to do with ability, whereas the word *may* has to do with having permission.

Compare with/compare to: Use *compare with* when examining or discussing similarities or differences. In general, only use *compare to* when representing a metaphorical similarity.

Complement/compliment: A *complement* is something that completes or brings to a whole. A *compliment* is an expression of congratulations or praise.

Comprise/compose: *Comprise* means to consist of or to include. *Compose* means to make up the constituent parts of, to constitute or form. Good examples are "The Union comprises 50 states" and "Fifty states compose (or constitute) the Union."

Continual/continuous: *Continual* means with occasional interruption, *continuous* means without interruption.

Data/datum: *Data* are plural, typically representing facts or information. *Datum* is the singular form of data, often used in the context of a point from which to measure.

Due to/because of: *Due to* means attributable to. *Because of* relates to a cause or reason for occurring. A helpful hint is that a sentence should not start with *due to*.

Effect/affect: An *effect* is a noun; it implies a result, something that is caused. *Affect* is a verb; it brings about a change. To affect is thus to influence or impress.

Either/neither: It is correct to write "either A or B" and likewise "neither A nor B," but we do not use "neither A or B." Moreover, in each case, these words imply only two options, hence we cannot say "either A, B, or C."

Essential/important: *Essential* implies indispensable, fundamental, or absolute. *Important* merely implies significant or noteworthy.

Farther/further: *Farther* should be used when the context is distance. *Further* implies something in addition, such as the need for further experiments. Hence, one does not move a fixture further toward the center.

Good/well: In most cases, *good* is used to modify a noun (e.g., she is a good writer), whereas *well* is used to modify a verb (e.g., she writes well).

However/nevertheless: Strunk and White (1979) suggest that we should avoid beginning a sentence with the word *however* when the meaning is *nevertheless* or *yet*. This is easily corrected via replacement with these more acceptable beginning words or by moving the *however* to the middle or end of the sentence. When used at the beginning of a sentence, *however* should be thought of as "in whatever way" or "to whatever extent." A good example is given by Strunk and White, (1979, p. 49): "However discouraging the prospect, he never lost heart." By contrast, it would be better to write "Nevertheless, he never lost heart despite the discouraging prospects" rather than to write "However, he never lost heart despite the discouraging prospects."

Imply/infer: Imply means to suggest or indicate by logical necessity, whereas *infer* means to deduce based on available evidence.

Precede/proceed: Precede means to come before in time, to occur prior to. *Proceed* means to go forward, especially after an interruption, or to move on in an orderly fashion.

Principal/principle: A researcher may be the *principal investigator* on the project but not the *principle investigator*. One may use a scientific principle but not a scientific principal. Another usage that is often confused is that the solution to an eigenvalue problem yields a principal value and in mechanics one may compute a principal stress or strain.

Shall/will: It is suggested by some that *shall* should be used for future expectations in first person and *will* should be used in second and third person. This distinction between *shall* and *will* occurs only in formal writing, however, and the word *will* often suffices. A good example is that *will* is appropriate in grant proposals, for example, "We will test the hypothesis that…"

That/which: In general, use *that* to lead into a defining or essential clause and use *which* to lead into an inessential or nonrestrictive clause. Kilpatrick (1984) suggests an easy way to decide usage in most cases: use *which* whenever the clause is set apart by commas and use *that* otherwise. The key point, however, is the word *that* is used with essential clauses. For example, note the difference between the following sentences. "The transducer that is broken is on the shelf." "The transducer, which is broken, is on the self." In the first case, only the transducer that is broken is on the shelf. In the second case, the transducer is on the shelf and it happens to be broken.

That/who: That refers to things and *who* refers to people.

While/whereas: Strictly speaking, *while* should be used to convey a sense of time, for example, "The computer acquired data while the device subjected the cells to increasing mechanical loads." Nonetheless, many accept *while* as a substitute for *although*. By contrast, *whereas* means "it being the fact that" or "inasmuch as."

Next, consider a few words that are useful in technical writing but sometimes misused.

Aforementioned: This word is an adjective; it must be combined with a previously used noun. For example, it is correct to write "The aforementioned finding suggests...," but it is incorrect to write, "As aforementioned, ..."

And/or: This is a construction used by some, but often best avoided. Use either the word *and* or the word *or* as appropriate.

Correlate: To put into a complementary, parallel, or reciprocal relationship, not implying causality.

Dilemma: Either a situation that requires one to choose between two equally viable alternatives or a predicament that seemingly defies a satisfactory solution.

Former: The first mentioned of two things.

Latter: Like *former*, this word implies two choices. If one has a list of three or more items, then to refer to the last one in the list, simply say "the last one," not "the latter one."

Per: "Pursuant to"

This: For clarity, follow the word *this* with a noun. For example, do not write, "This is to be expected," but rather write, "This nonlinearity is to be expected" or "This finding is to be expected."

Finally, some words have specific meanings in science and mathematics even though they are often used loosely in everyday speech. Because our interest is scientific writing, we must respect the specific meanings. Three prime examples of such words are *significant*, *necessary*, and *sufficient*. It would be natural, for example, to write: "The response of Group A differed significantly from that of Group B." Yet, we must ask whether this is what we really mean. The word *significant* in science usually carries a statistical meaning, that is, it usually implies a finding based on a standard statistical test, that there is a significant difference between two metrics (e.g., as indicated by a $P < 0.05$ associated with a specific statistical analysis). If such a test was performed and passed, then we could write our illustrative sentence as given; if not, it would be better to use a different word or to delete the modifier altogether. In the absence of a statistical test, it would be better to write: "The response of Group A differed markedly from that of Group B" or simply "The response of Group A differed from that of Group B." The words *necessary* and *sufficient* similarly have precise meaning in mathematics and they are often used together. In this context, *necessary* means required and *sufficient* means adequate. For example, a necessary and sufficient condition for a solution to hold is much stronger than a sufficient condition alone.

In concluding this section, it is interesting to consider a comment ascribed to the aforementioned famous ancient philosopher Socrates:

The wise man knows that he knows not; the fool knows not that he knows not.

Similarly, consider a comment by a famous modern philosopher Bertrand Russell (1872–1970):

Although this may seem a paradox, all exact science is dominated by the idea of approximation....When a man tells you he knows the exact truth about anything, you are safe in inferring that he is an inexact man.

If we accept that science represents relative, not absolute, truth, then we should be careful when using strong phrases such as "the data demonstrate" or "the data prove." For example, consider alternative phrases such as "the data suggest" or "the data imply." Similarly, should we not avoid saying that something "is," but instead say that "it appears that." Here, again, we suggest that one should think carefully about this issue, make a purposeful decision within the context of common usage, and be consistent.

Exercise 2.8
Read three technical papers in your field and generate a list of phrases that reflect either a "certainty" or a "possibility" with regard to important findings or conclusions. Write a one-page summary and indicate the approach to communicating such results that you find to be the best.

2.6   Punctuation, Abbreviations, and Foreign Languages

2.6.1   Exploit Methods of Punctuation

Punctuation is a system of devices or marks (e.g., commas, semicolons, colons, dashes, and parentheses) that clarify relationships between words and groups of words (Vivian & Jackson, 1961). Aside from the standard use of the period, many writers of science and engineering tend to use commas sparingly and avoid using semicolons, colons, and dashes. Although we should not overuse such devices, variety in punctuation can be as effective in written communication as variety in tone can be in oral communication. We list here a few rules of punctuation, but encourage the reader to give particularly careful thought to the effective use of semicolons, dashes, and parentheses. As a start, consider Rules 2–4 of Strunk and White (1979).

Traditionally, it has been suggested that one should use a comma after each entry, except the last, in a list of three or more entries that share a common conjunction such as *and* or *or*. For example, write "this finding was unexpected, repeatable, and important." To appreciate this usage, recall from Sect. 2.1 that the fourth step in writing well is "read out loud." Doing so here, the ear reveals a difference between "this finding was unexpected, repeatable, and important" (with a verbal pause after each comma) and "this finding was unexpected, repeatable and important." In other words, the latter case sounds like the finding was both "unexpected" and "repeatable and important." Yet, this usage of the "serial comma" now differs by publisher and thus one should seek to be consistent with common usage.

When paired, commas are useful devices to set off a nonessential clause, for example, "The transducer, which is broken, is on the shelf." When used with a conjunction to introduce an independent clause, the comma should be omitted before the *and* when the clauses relate closely. By contrast, a comma should almost always precede conjunctions such as *but*, *for*, and *or*. Commas are also useful to set off an introductory phrase, such as "In this paper, we show..." Finally, a comma can be used to separate three or more modifiers, such as in the case of a "randomized, double-blind, clinical trial."

Use the semicolon instead of a period when independent clauses relate closely and it is effective to highlight this similarity. The only exception to this rule is the case of short independent clauses. Consider, therefore, the following examples from Strunk and White (1979):

Stevenson's romances are entertaining. They are full of exciting adventures	Stevenson's romances are entertaining; they are full of exciting adventures
It is nearly half past five. We cannot reach town before dark	It is nearly half past five; we cannot reach town before dark
Man proposes, God disposes	– no change
Here today, gone tomorrow	– no change

The semicolon is also useful to separate main clauses that are joined by conjunctive adverbs such as the following: *indeed*, *yet*, *however*, *moreover*, or *hence*. For example, we might write (Iverson et al., 1998): "This consideration is important in any research; yet it is often overlooked, if not denied."

Use the colon before a long in-line quotation (see Sect. 2.7), to introduce a list, or to separate independent clauses when the first clause introduces the second one. For example, if we wish to specify the composition of a physiological solution used in an experiment, we might write the following.

The specimens were immersed in a physiological solution consisting of, in mM: 116.5 NaCl, 22.5 Na_2HCO_3, 1.2 NaH_2PO_4, 2.4 Na_2SO_4, 4.5 KCl, 1.2 $MgCl_2$, 2.5 $CaCl_2$, and 5.6 dextrose.

Like the comma, one can use short dashes (or em dashes) and parentheses to set off nonessential, but clarifying, clauses or entries. The decision to use the em dash or parenthesis (more common) is again a matter of style, with the em dash typically reserved for the longer, sometimes tangential, breaks in thought. Consider the following two cases:

Of the many risk factors for coronary artery disease—high cholesterol, high salt intake, cigarette smoking, lack of exercise, diabetes, and hypertension—some can be avoided by simple changes in lifestyle.

Many risk factors for coronary artery disease can be controlled by simple changes in lifestyle (e.g., high cholesterol, high salt, and smoking).

Parenthetical setoffs are also useful in providing supplementary information or identifications. For example, it is common to read: "of the 10 tests, only 5 (50%) were successful," "the differences were not significant ($P > 0.05$)," or "a consistent volume of fluid (10 ml) was injected." As noted below, it is common to include clarifiers within parenthetical set offs such as (for example, ...) or (that is, ...), abbreviations for which are given below.

The hyphen has many uses as well; see Brogan (1973) for a good discussion of this device. Commonly misunderstood uses are numerical or multiword modifiers. For example, we should write "The diameter of the device is 5 mm," but we should use the hyphen to write "The 5-mm diameter device..." We should also use the hyphen to write out numbers such as thirty-seven or two-thirds. Another use of a hyphen is in the pairing of words that, via the natural evolution of grammar, often become single words. A simple example is *mechanical transduction*, which became *mechano-transduction* and now *mechanotransduction*. Finally, multiword modifiers are often hyphenated, for example, "the signal-to-noise ratio" or "one-way Student *t*-test." Uses such as "pre- and post-surgical" are also common.

With regard to numbers, it is common to write out in words those numbers less than 10 (e.g., zero, one, two) but to write out numerically those numbers 10 or greater (e.g., 11, 100, 1000). There are exceptions, however (Blake & Bly, 1993). For example, if data are collected at days 0, 3, 7, and 14, we would not write "at days zero, three, seven, and 14." In other words, one of the best rules of thumb is consistency for clarity. Moreover, always write large numbers in a way that is most easily understood. For example, the number 30,000,000 may be best understood as 30 million if referring to dollars or population; by contrast, it may be best understood as 3×10^7 if referring to a quantity in physics or chemistry or 30 MPa if referring to stress in mechanics. The example of 30 MPa reminds us to use, when appropriate, accepted prefixes: giga (G), mega (M), kilo (k), milli (m), nano (n), and so forth. The best rule of thumb, therefore, is always write for clarity to an intended audience. Finally, it should be noted that decimal values less than unity should be written with a leading zero, for example, 0.15 rather than .15. Whether one uses decimal values or not, also remember to include only *significant digits*, that is, information that is reliable. For example, although a calculator or computer may provide an answer of 4.1248432, but if only three of the digits are reliable we should write this as 4.12. Refer to elementary textbooks on physics or chemistry for good discussions on the appropriate use of significant digits.

As last reminders, do not use the apostrophe in special cases of decades or centuries, rather one should write 1970s or 1700s. Words such as *its* and *it's* and *whose* and *who's* are often confused. This is simple to remember: *it's* and *who's* are contractions of "it is" and "who is," whereas *its* and *whose* show possession. Contractions should be avoided in formal writing, however. Finally, it is important to emphasize that most publishers use a single space after a period, not two spaces. Not only is the single space typically more pleasing to the eye, it is also an effective means to reduce the number of pages and hence the cost of a print publication because those extra spaces add up.

2.6.2 Abbreviations

Many writers suggest that abbreviations should be avoided in formal writing. In technical writing, however, we should merely minimize the use of abbreviations, using them only when they improve conciseness or are common within the intended context. One of the easiest ways to decide whether to use a particular abbreviation is to ask if it will improve or impede a reader's understanding. For example, many readers of technical papers go directly to the figures or Results to see what was found or they go directly to the Discussion to see what was deemed to be important. They can be frustrated if the figure legends, Results or Discussion contain uncommon abbreviations that require them to search the Introduction or Methods to find the associated meanings. This should be avoided. By contrast, many abbreviations are so common that they can often be used without explanation. Examples include: ANOVA (analysis of variance), DNA (deoxyribonucleic acid), MRI (magnetic resonance imaging), and mRNA (messenger ribonucleic acid). Scientific units can also be abbreviated without definition, as, for example, kPa (kilopascal), MHz (megahertz), ml (milliliters), mmHg (millimeters of mercury), and mM (millimolar).

Many other abbreviations, such as LV (left ventricle) or MAP (mean arterial pressure), are used widely, but must be defined. This is also the case for abbreviations of many biologically important molecules and chemical compounds. For example, one would be expected to define and use the following abbreviations: NaCl (sodium chloride), NO (nitric oxide), TGF-β (transforming growth factor-beta), and PMMA [poly(methyl methacrylate)]. In these cases, however, common practice is to introduce the abbreviation only if it is used three or more times in subsequent text and to define the abbreviation at its first occurrence in the body of the paper, not in the Abstract. It is best not to construct new abbreviations, however, just because a descriptor is used repeatedly. For example, we would not introduce NM for noun modifier even if used extensively. Similarly, we would not use SI for split infinitive. Indeed, this example reveals that one should be careful not to define new abbreviations that are identical to commonly accepted abbreviations (e.g., SI stands for *Systeme Internationale*, the common units of measurement in most of science and engineering). Iverson et al. (1998) provides an extensive list of accepted abbreviations in medicine.

2.6.3 Foreign Languages

Many publishers seek to reduce the length of a publication because additional printed pages translate into additional costs. For this reason, some well-accepted abbreviations are encouraged and thus are common. Four of the most common abbreviations come from Latin, namely:

1. et al., which means "and others," is commonly used when referring to a publication by three or more authors. In such cases, it is customary to cite the last name of the first author followed by "and others," for example, Smith et al. (2021) or (Smith et al., 2021). Whether one italicizes the Latin et al. depends on the publisher, but in most cases, a period should follow the *al* alone.
2. e.g., which means "for example," is often used in parenthetical situations (e.g., in this way). Remember, too, that an example is just that, one of many possible illustrations; it is not a unique clarifier.
3. i.e., which means "that is," is also often used in parenthetical situations (i.e., it often appears in this type of context). In contrast to e.g., using i.e. is similar to using the phrase "in other words" and thus is meant to clarify a meaning, not to provide an illustrative example.
4. cf., which means "compare with," is often used to draw attention to a similar or related illustration, equation, or other scholarly work, as, for example, (cf. Fig. 1).
   Another common abbreviation from the Latin is,
5. etc., which means "and other unspecified things of the same class" or simply "and so forth." Albeit commonly used, most grammarians agree that etc. should not be used in formal writing or, if so, only sparingly for good purpose. If one does not wish to provide an exhaustive list, using "for example" is an appropriate way to indicate the listing of some, but not all, members of that class. One would thus never use e.g. and etc. within the same parenthetical statement.

Although scientific and engineering documents should be scholarly, they should not be pretentious. An attempt to impress the reader by using phrases or words from Latin, Greek, or other "foreign" languages is not advised in general unless their meaning is well known and they engender conciseness or clarity. For example, some words and phrases are common in the biomedical literature and should be used, for they are well understood. In addition to the aforementioned et al., i.e., e.g., and cf., consider for example:

de novo: a new
in situ: in its natural place
in vitro: in glass or generally in an artificial environment
in vivo: within a living organism
ex vivo: outside of a living organism, but still living.

Other acceptable, but less common, examples are:

ad infinitum: without end or limit
in toto: totally, altogether, entirely
reductio ad absurdum: reduction to the absurd
status quo: as it is now.

2.7 Footnotes, Quotations, and Proper Citation

2.7.1 Footnotes

Footnotes are brief notes placed at the bottom of a page that provide a citation (older use) or a comment on a specific part of the main text. Although scientists and engineers used footnotes extensively in the past, such usage is generally discouraged today. We do not advocate eliminating footnotes, but we encourage sparse usage. For example, footnotes can provide brief examples or clarifications that do not otherwise fit within the text using parenthetical devices such as commas, parentheses, or em dashes. Footnotes should not be used to solve problems in organizing material or sentence structure, however.

2.7.2 Quotations

Quotations must be denoted in one of two ways: if integrated within the text, they must be enclosed within quotation ("") marks; if longer, and singled out, they should be indented but appear without quotation marks. Some publishers also use a smaller font for indented quotations. For example, let us recall the quotation from W. Strunk Jr. that is given at the beginning of this chapter:

> Vigorous writing is concise. A sentence should contain no unnecessary words, a paragraph no unnecessary sentences, for the same reason that a drawing should have no unnecessary lines and a machine no unnecessary parts. This requires not that the writer make all his sentences short, or that he avoid all detail and treat his subject only in outline, but that every word tell.

In general, use longer block quotations sparingly, if at all, in a technical document. Many times, the reader will skip such quotations to get to the meaning or importance of the quotation that follows. Another rule of thumb is to ask whether the quotation is necessary or if it is simply an easier alternative. If the latter, a brief reference to the original source material followed by original commentary would be better.

Ellipses, that is, three dots in sequence (...), indicate that words are omitted, usually from a quotation. Four such dots in sequence usually indicate that words are omitted at the end of a sentence, hence the last dot can be thought of as the period at the end of that sentence. Beginning a quotation with a lowercase letter indicates that the author has omitted the initial part of the quote; beginning with a capital letter indicates that one is beginning the quotation at the beginning of the sentence. Thus, ellipses are not needed at the beginning of a quotation.

When information is missing or incorrect in a quotation, it is appropriate to provide complete and accurate information. The information that is added should be enclosed by brackets [ ]. For example, given the quote "Newton postulated...," one may write "[Sir. Isaac] Newton postulated..." It is acceptable, however, to provide a quotation exactly as it appears without correcting obvious or subtle errors provided this does not mislead the reader. Finally, one may find or insert [sic] in a quotation. The form [sic] comes from the Latin and indicates that a seemingly paradoxical word, phrase, or fact is not a mistake; it should be read as given.

2.7.3 Proper Citation

Although we discuss issues of ethics in Chaps. 6–8, including plagiarism, we briefly mention it here as well. Simply put, plagiarism is often defined as passing off as one's own the ideas and words of another. Actually, "pass off" is too soft; plagiarism is intellectual theft. Most universities have writing centers and associated Web pages, hence one can find formal definitions and excellent examples of plagiarism.

The best way to avoid plagiarism is through proper citation. Although we tend to learn in English classes that there is a particular way to cite works in a bibliography, the citation format varies considerably from publisher to publisher in scientific writing. Hence, the best advice is to read the *Instructions for Authors*, which can be found on the website of the journal or publisher or often within the journal itself. We give examples of different styles of citation in Sect. 3.1.9.

> **Exercise 2.9**
> Investigate a famous case of plagiarism in science or engineering and write a concise three-page report.

2.8   Vocabulary

Vigorous writing should be clear and concise; nevertheless, it should also be provocative and engaging. The reader is thus encouraged to read Chap. 5 in Strunk and White's *The Elements of Style.* One should employ words of power (i.e., having strong meaning) without becoming verbose or haughty. One way to accomplish this is to expand our vocabulary, which is perhaps best accomplished by keeping a diary of effective and engaging words as we read technical papers and books. When we come upon a forceful, precise, or attractive word, we should take note of it. Knowing that the author may have misused the word, however, we should always consult a reliable dictionary when recording the associated definition. A good dictionary can be found online at www.m-w.com. Here, we list a few words that one can use advantageously in technical writing, which may or may not be used on a daily basis by the reader.[4]

Abate: To reduce in amount, degree, or intensity; lessen.
Adverse: Antagonistic in design or effect; hostile; opposed.
Ancillary: Subordinate.
Assume: To take for granted; suppose.
Attenuate: To make slender, fine, or small.
Augment: To make greater, as in size, extent, or quantity; enlarge.
Causal: Pertaining to or involving a cause.
Caveat: A warning or caution.
Cogent: Forcibly convincing.
Copious: Yielding or containing plenty; affording ample supply.
Corroborate: To strengthen or support; attest the truth or accuracy of.
Cull: To pick out from others; select.
Cursory: Hasty and superficial; not thorough.
Delve: To search deeply and laboriously.
Didactic: Intended to instruct; expository.
Disparate: Completely distinct or different in kind; entirely dissimilar.
Dubious: Fraught with uncertainty or doubt; uncertain.
Egregious: Outstandingly bad; blatant; outrageous.
Eminent: Towering above others; projecting; prominent.

---

[4] These definitions are taken largely from the American Heritage Dictionary.

Enigma: An obscure riddle; puzzling, ambiguous, or inexplicable.

Equivocal: Capable of two interpretations; cryptic; evasive; ambiguous.

Erudite: Deeply learned.

Exacerbate: To increase the severity of; aggravate.

Extant: Still in existence; not destroyed, lost, or extinct.

Extenuate: To lessen or attempt to lessen the magnitude or strength of.

Fortuitous: Happening by accident or chance; unplanned.

Fraught: Attended; accompanied.

Glean: To collect bit by bit.

Hypothesize: To assert a hypothesis (i.e., an assertion subject to proof).

Inadvertent: Accidental; unintentional.

Inchoate: In an initial or early stage; just beginning; incipient.

Inexplicable: Not possible to explain.

Inordinate: Exceeding reasonable limits; immoderate; unrestrained.

Integral: Essential for completion; necessary to the whole.

Intrinsic: Pertaining to the essential nature of a thing; inherent.

Lucid: Easily understood; clear.

Manifold: Of many kinds; varied; multiple.

Marked: Having a noticeable character or trait; distinctive; clearly defined.

Myriad: A vast number; a great multitude.

Nadir: The place or time of deepest depression; lowest point.

Nullify: To make ineffective or useless.

Obviate: To prevent or dispose of effectively; to render unnecessary.

Ostensible: Given or appearing as such; seeming; professed.

Ought: Indicates obligation or duty, prudence, or desirability. Use with *to*.

Paucity: Smallness of number; fewness.

Permeate: To spread or flow throughout; pervade.

Peruse: To read or examine, especially with great care.

Posit: To put forward as a fact or truth; to postulate.

Postulate: Something assumed without proof as being self-evident or generally accepted, especially when used as a basis for an argument.

Premise: A proposition upon which an argument is based or from which a conclusion is drawn.

Proliferate: To reproduce or produce new growth rapidly and repeatedly.

Promulgate: To make known by public declaration; announce officially.

Propensity: An innate inclination; tendency; bent.

Purview: The extent or range of function, power, or competence; scope.

Quiescent: Inactive or still; dormant.

Recant: A formal retraction of a previously held belief or statement.

Recondite: Not easily understood by the average person.

Reiterate: To say over again.

Replete: Plentifully supplied; abounding.

Requisite: Required; absolutely needed; essential.

Retrospect: A review, survey, or contemplation of things in the past.

Salient: Striking; conspicuous.

Spurious: Lacking authenticity or validity; counterfeit; false.

Substantiate: To support with proof or evidence; verify.

Succinct: Clearly expressed in few words; concise; terse.

Sundry: Various; several; miscellaneous.

Surmise: To infer without sufficiently conclusive evidence.

Tacit: Not spoken.

Tantamount: Equivalent in effect or value. Used with *to.*

Tractable: Easily managed or controlled; governable.

Ubiquitous: Seeming to be everywhere at the same time; omnipresent.

The space below allows you to record additional words, with their definitions, that you would like to add to your technical vocabulary.

2.9   Closure

Recalling that one of the best ways to improve one's writing is to read widely, we should not only read for pure enjoyment or the gaining of new technical information, we should also read with the intent of learning how to write well. In other words, take note of the effectiveness of active versus passive voice, the tense, the infinitives, the effective use of punctuation marks, and so forth; record and use particularly forceful words and phrases. Aside from scientific publications, works of history, philosophy, and theology (one of the four original academic disciplines, with law, medicine, and natural philosophy) often present good examples of writing well.

As a specific example, consider the book *On Growth and Form* by D'Arcy Thompson (1917). In the foreword of the 1961 abridged edition, P.B. Medawar wrote that Thompson's work was "beyond comparison the finest work of literature in all the annals of science that have been recorded in the English tongue." What gave rise to such a claim? Thompson was a true scholar, with expertise in the classics, mathematics, and zoology; moreover, he purposed both to document good science and to write well. Although we should not expect to achieve such success in writing well, we should remain committed to producing the best work possible.

References

Berry, T. E. (1971). *The Most Common Mistakes in English Usage*. McGraw-Hill.

Blake, G., & Bly, R. W. (1993). *The Elements of Technical Writing*. Macmillan.

Brogan, J. A. (1973). *Clear Technical Writing*. McGraw-Hill.

Clendening, L. (1960). *Source Book of Medical History*. Dover Publications.

Day, R. A., & Gastel, B. (2006). *How to Write and Publish a Scientific Paper* (6th ed.). Greenwood Press.

Gibaldi, J. (1995). *MLA Handbook for Writers of Research Papers* (4th ed.). The Modern Language Association of America.

Iverson, C., et al. (1998). *AMA Manual of Style: A Guide for Authors and Editors* (9th ed.). Lippincott Williams & Wilkins.

Kilpatrick, J. J. (1984) *The Writer's Art*. Andrews and McMeel, Kansas City.

Strunk, W., & White, E. B. (1979). *The Elements of Style* (3rd ed.). Allyn and Bacon.

Valiela, I. (2001). *Doing Science: Design, Analysis, and Communication of Scientific Research*. Oxford University Press.

Vivian, C. H., & Jackson, B. M. (1961) *English Composition*. Barnes & Noble.

Scientific Publications

3

3.1 Basic Content

Different types of publications in science and engineering include abstracts, conference proceedings, journal articles, books, dissertations, and technical reports. We focus, however, on that which is generally regarded as most important, the archival journal article. There are also different types of journal articles, including original articles, technical notes (sometimes called brief communications), and expert reviews. We focus on the original article, which is both most common and most important to the advancement of science and engineering for it details novel approaches and findings. Some journals impose stringent guidelines on the organization of such articles, yet considerable flexibility often allows the author(s) to present the material in ways that reflect personal style and yet maintain standards and conventions. For purposes of illustration, consider two common outlines recommended by the majority of scientific journals, either

Abstract (plus keywords)
Introduction
Methods (or Materials and Methods)
Results
Discussion
Acknowledgments and Disclosures
References

or

Abstract (plus keywords)
Introduction

J. D. Humphrey and J. W. Holmes, *Style and Ethics of Communication in Science and Engineering*, Synthesis Lectures on Engineering, Science, and Technology, https://doi.org/10.1007/978-3-031-39125-5_3

Results
Discussion
Methods (or Materials and Methods)
Acknowledgments and Disclosures
References

In the former, it is assumed that it is critical first to understand the (often novel) methods that enabled the findings; in the latter, it is assumed that first enumerating the (often well accepted) methods can distract from the central contribution of the paper, namely, the results. Regardless, it is increasingly recognized that the utility and potential importance of any results depend fundamentally on rigorous and reproducible methods, which will be discussed in detail below and relates to transparency.

Recall from Chap. 2 that there are five basic steps of writing well: formulating a detailed outline, writing freely, editing critically, reading out loud, and having a colleague critique the final draft. A detailed outline not only includes major headings, as noted above for the original journal article, it also includes potential subheadings and either bullets that highlight key ideas or in some cases leading sentences. Although different authors employ different approaches to craft a detailed outline, a good place to start is to call all authors together and to discuss the primary findings: figures, images, tables, equations, and so forth. These key findings can then be discussed and ordered logically, which will define the key bullets in Results and serve to remind us what methods were essential in obtaining the results. Once we have outlined the Results and Methods, it is then easier to outline the Discussion and Introduction and to begin the process of composition. We discuss each of these, and other, key sections in detail below.

3.1.1 Results

This section is the heart of the technical paper; it reports the primary findings, which should represent the most important contribution of the paper. The results section should be relatively easy to write, thus many authors prefer to start here. Indeed, in cases of multiple authors working on a single document, following appropriate discussion amongst the authors, the first author often drafts the Results and Methods, then the senior author drafts the Discussion and Introduction. All authors then revise the completed first draft. Regardless of approach, consistent with the above, one of the best ways to write Results is to collect together all of the figures, tables, equations, or other major findings that you may include, then prioritize and order them accordingly. It is important to emphasize that we need not order the results chronologically; rather, order the findings such that the narrative flows logically. A key challenge in this regard is to help the reader to understand within a short period that which took you weeks or months to glean from the findings, whether experimental, computational, or theoretical. Once outlined, it is then easier to write freely. Some journals require subheadings within the results while others do not. In the former, there is a trend towards using declarative statements as subheadings (e.g., *Microscopy revealed...* rather than *Microscopic Findings*). In the latter, the lead sentence of each paragraph typically serves to introduce each key finding. Indeed, some recommend that the lead sentence of each paragraph in Results should state the most notable finding reported in that paragraph.

One of the most frequently asked, and often most difficult to answer, questions is: How much interpretation of the findings should be in Results versus Discussion? One reason that this is difficult to address is that it depends in part on the style of the author and recommendations by the specific journal. Indeed, some journals require results and discussion to be synthesized within one section. In general, however, most technical writers agree that the results section should be objective; it should focus on presenting the findings. Hence, although it is common to point out within Results any interesting or important features within specific figures, images, equations, or tables, it is generally best to reserve for the Discussion any interpretation of the significance of the findings as well as any comparison to findings by others. Given the diversity of styles and publications now available, however, the best place to start this thought process is by reading the publication-specific *Instructions to Authors*.

Another question that often arises is how best to refer to a figure or table. For example, should we write

A and B were found to be related linearly (Fig. 1),

or is it better to write

Figure 1 reveals a linear relationship between A and B.

In other words, is it best to state the key finding and refer parenthetically to the associated figure, image, equation, or table, or is it best to cite directly the particular evidence that reveals (shows, illustrates, or so forth) the key finding? Notwithstanding some exceptions (e.g., specific *Instructions for Authors* found on a journal's website), the answer to this question is often that it is a matter of personal style. Note from our illustrative example, however, that the first approach involves passive voice whereas the second approach involves active voice but further requires the figure, image, equation, or table to "do something"—to reveal, show, illustrate, confirm, or so forth. As noted previously, some editors suggest that inanimate devices such as figures cannot reveal or show such things, thus they prefer parenthetical references over the direct approach. Conversely, others prefer the crisp, active voice in the second example, which helps to minimize the use of passive voice as desired in general. We encourage the reader to consider these and similar options carefully and to adopt a consistent, but not rigid, personal style within the allowances of the particular journal. Such a decision should affect other aspects of writing a technical paper, for example, in the Introduction wherein one often reads "This paper presents." Again, some would argue that only the investigators can present, not the paper, yet many prefer this crisp, active style of introduction. Of course, also recall from Chap. 2 that one can often convert a sentence from passive to active voice in others ways, as, for example, "A and B were found to be related linearly" could be rewritten to read "A and B related linearly."

3.1.2   Methods (or Materials and Methods)

Regardless of its placement within the submitted paper, the section on methods is usually the easiest to write. Indeed, if one has trouble getting started in the "write freely" phase, it is often best to go to Methods. Simply put, this section is where we document how we accomplished the work. In principle, we should provide enough detail to allow the reader to repeat the study in the same way—this alone allows one to test and confirm a basic tenet of science, reproducibility. Noting that many recent changes in scientific publishing seek to increase rigor and reproducibility, this is discussed further in Sect. 3.1.7. Here, simply note that the increasingly sophisticated methods and procedures used in science and engineering demand significant planning both in designing and describing the research. Given that one can often use commercially available kits or software packages, however, there is also a need for balance between detail and proper citation.

Two effective devices in writing the methods section are to use ample subheadings and parallelism. Here we emphasize that scientific papers should not be written to be read, but rather to be studied. Subheadings thus aid the reader in organizing information or quickly locating particular aspects of the methods when needed. Typical subheadings in a paper on cell biology could be:

Immunohistochemistry
Confocal Microscopy
Statistics

Subheadings in a paper on biomechanical modeling could include:

Theoretical Framework
Constitutive Models
Numerical Methods

In either case, subheadings should proceed logically and thereby reveal to the reader the thought process or workflow followed by the investigators. The format for the subheadings, for example, numbering or italicizing, is dictated by the particular journal and thus is provided in the specific *Instructions for Authors*.

Because many scientific findings result from or imply a mathematical statement, it is important to address the treatment of equations within a paper, often within Methods or Results. Simply put, write an equation as a normal part of a sentence. For example, Newton's famous second law of motion is usually written as $f = ma$, where f denotes force, m denotes mass, and a denotes acceleration. Consistent with the presentation here, mathematical symbols usually appear in a distinct font, which may include italics (e.g., scalars) and boldface, or both (e.g., vectors as boldface italics and tensors as boldface non-italics). In many cases, either for emphasis or simply because of complexity, equations are set off as separate lines within the document. In this case, the equation is still part of a sentence and thus should include commas and periods as appropriate. For example, one could write the following:

It is important to remember that Newton's second law of motion, namely

$$f = ma,$$

holds only with respect to an inertial frame of reference.

Similarly, we could write the following:

The governing equation in this case is Newton's second law of motion, which can be written as

$$f = ma.$$

An easy way to remember that equations are part of the normal grammatical structure is to recall step 4 from Chap. 2 on how to write well—reading out loud forces us to include equations as natural parts of the text. A final reminder is that most journals do not allow a nomenclature section for symbols used within equations. Hence, one should state the meaning of a symbol just before or just after the equation in which it is introduced, as we did above within the in-line example for Newton's second law (e.g., noting that the vector f denotes force, the scalar m denotes mass, and the vector a denotes acceleration).

There is no need to define symbols that are universally accepted or familiar to readers of a particular journal. Examples of well-recognized symbols include those for summation, derivatives, integrals, and so forth. Given the increasing inter-disciplinary nature of science and engineering, however, commonly used variables could represent multiple quantities within the same paper, hence there is a need for care. For example, R could be used for radius, but it could also be used for the universal gas constant [$R = 8.314$ J/(g mol) K] or a coefficient of determination (R^2). The most important suggestion in this regard is to be clear and self-consistent, and to eliminate redundancies when possible (e.g., use r for radius if using R for the universal gas constant).

Finally, a frequently asked question relates to the level of detail needed in cases where one reports results obtained using methods reported previously in other journal articles. Although there is no definitive answer, a reasonable practice is to document the essential new methods and refer the reader to previously established methods, where appropriate. For example, if your group established the previous methods, state that the details can be found in a previous publication, then briefly outline the methods; if others established the previous methods, cite the key paper(s) and provide a brief, but slightly more detailed, summary of the methods. Conversely, one must provide significant detail when reporting a new method or procedure. Such detail should include specific instruments and vendors, chemicals and their sources and concentrations, specific versions of software, and so forth. Toward this end, some journals now require Supplemental Materials that include a detailed "Major Resource Table" wherein vendor numbers, concentrations, dilutions, and so forth are given precisely. Again, providing such detail helps to promote rigor and reproducibility. In cases of human or animal research, one must also document that the appropriate institutional oversight committee (e.g., the Institutional Review Board, or IRB, for human research or the Institutional Animal Care and Use Committee, or IACUC, for animal research) approved the work reported.

Although the reader is referred to Sect. 7.5 for additional discussion, the increasing emphasis on detailed reporting of methodology has created another challenge for authors. A basic tenet of science is consistency—in principle, the only thing that should change in a sequence of experiments is the sample itself and the only thing that should change in a sequence of computations is the constitutive relation or boundary/initial conditions. That is, a well-designed sequence of experiments or computations (which may be reported via multiple papers) should use the exact same methods, which increases reproducibility. In such cases, the description of such methods necessarily should be identical in a detailed Methods section. Yet, while asking for extreme detail, many editors and publishers also discourage or disallow one to use verbatim the description of methods from prior papers—this is often termed self-plagiarism.

3.1.3 Discussion and Conclusion

Most journals recommend against using a separate section for conclusions, which would typically be brief, hence Discussion often serves a dual role. One should address at least three points in this section of the paper:

Interpret the specific results and emphasize the significance.
Compare the current with past results.
Identify limitations and potential needs for further research.

Whereas the Introduction normally addresses the significance of the overall research topic or area, the Discussion should address the potential significance of the particular findings. For example, an Introduction may note the importance of cardiovascular consequences of hypertension, which affects more than 100 million Americans, but the Discussion may note the significance of the new finding that blocking a particular receptor in particular cells reduces hypertension in an experimental cohort. Like the Introduction, therefore, the Discussion should cite appropriate references, albeit often with greater elaboration of the relevant details. It is important in this regard to cite only the most relevant literature. In other words, the goal is to place the current findings within the most appropriate context, not to provide an exhaustive collection of all previous work that relates remotely to the overall topic or specific findings. Because of the explosion of scientific and engineering knowledge, citing authoritative review papers can often serve to cover general information without concern that some important papers may be missed. Given that Review Articles tend to be highly cited, many journals now solicit such papers and there is now a plethora of Review Articles, not all of which are authoritative or insightful. Toward this end, authors should not cite Review Articles casually or simply because a particular review has been well cited; rather, authors should read different reviews and select those that provide the most relevant commentary and list of references. Related to issues of ethics, of course, one should not purposely fail to cite a relevant paper for personal gain.

A frequent question with regard to the Discussion is how much information should be included on the inherent limitations or future needs. In some ways, this addresses both the style and the ethics of written communication. It is both prudent and useful to others to point out limitations of the study, with justifications, for this will both put the current study in perspective and guide future work. Nevertheless, one must be careful not to focus on the negatives in a way that it distracts from the significant accomplishments or advancements of the study. The key, therefore, is to maintain an appropriate and candid balance. Similarly, it is useful to point the reader toward important directions for future research. Yet, many investigators avoid this for two reasons. First, there is concern that a reviewer may ask the authors to include the additional studies in a revised submission of the current work. Second, some authors prefer not to reveal ideas that they may later build on to achieve further advances. Again, it is useful to maintain a proper balance—provide guidance so that others can advance the field while protecting novel ideas and intellectual property.

In summary, the primary goals of the Discussion are to reemphasize the significance or innovation of the study, to interpret and discuss implications of the specific findings, to compare the current findings with similar work by others, to discuss limitations of the methods or findings, and when appropriate to provide direction for future work.

3.1.4   Introduction

As with any Introduction, the primary goal of this section is to capture the reader's interest and thereby "set the stage." Toward this end, it is generally recommended that the Introduction answer three basic questions:

Why is the general topic or particular study important?
What is currently known and what remains unknown?
What does the current paper address or accomplish?

One should be able to answer these questions easily after having written Results. Consistent with answering these questions, the typical Introduction consists of three to four paragraphs even though there is considerable variety in the number and especially the lengths of these paragraphs. Experienced writers may write the Introduction first, but most writers address the Introduction after completing the Methods and Results and sometimes even the Discussion. Regardless, it is important to provide sufficient references therein to justify both the need for the study and the general approach adopted.

An important issue with regard to writing a journal article, including the Introduction, is the appropriate use of abbreviations. Good rules of thumb are to use only commonly known abbreviations (e.g., DNA for deoxyribonucleic acid and MRI for magnetic resonance imaging are both common and appropriate whereas ROT for rules of thumb would not be useful or appropriate), to use them only if the word or phrase is repeated three or more times throughout the document, and to introduce them at the first occurrence in the body of the paper (cf. Sect. 2.6.2). Some journals require the author to collect abbreviations together, for example, in a footnote on the first page of the paper or in a table. Regardless, it is best to use abbreviations sparingly.

3.1.5   Abstract

The technical abstract has always served an important role—it provides a brief summary of a paper and thereby helps a reader to decide whether to read or study the paper. With the advent of on-line search engines, however, the Abstract has become a particularly important means of capturing the attention of the intended audience. Hence, albeit a short section, often not more than 250 words, the Abstract deserves great attention.

Most writers compose the Abstract last. It must reflect briefly the overall paper, including the basic motivation, significance, general approach, and key discoveries or final solution; it must be written clearly, without jargon, acronyms, or uncommon abbreviations, and must stand alone, typically without references. As with the Introduction, the first sentence of the abstract should be engaging. In contrast with the Introduction, the last sentence of the abstract either summarizes the most important finding or points to pressing needs for future research. Whereas most journals allow the authors to write the Abstract as they prefer, a few journals require the authors to follow a uniform outline, including specific subheadings. As in all cases of technical writing, it is thus important to read the *Instructions for Authors* for the particular journal.

3.1.6 Acknowledgments and Disclosures

It is customary, indeed required in most circumstances, to document any financial or related support that enabled the work from ideation to publication. Such support can include grants, contracts, use of core facilities, in-kind gifts, and support of salary from sources other than the authors' stated institutional affiliations. As an example, one might read:

> This research was supported by grants from the National Institutes of Health (R01 HL-100000 and R21 HL-010000).

In addition to financial support, it is appropriate to acknowledge technical support, editorial contributions, and advice by individuals who do not merit co-authorship because they did not contribute substantively to the intellectual content of the paper (see Chap. 6) but who contributed nonetheless. Such acknowledgment must be merited, however, and those named must be informed. Indeed, some journals now require individuals who are acknowledged in this section to stipulate in writing that they are both aware of and deserving of such recognition.

More recently, also in the interest of transparency, there has been a move to document the specific contributions of each author as well as any relevant disclosures or conflicts of interest, real or perceived. Author contributions are typically indicated by initials (e.g., JDH for Jay D. Humphrey or JWH for Jeffrey W. Holmes) and can include (e.g., for the publisher Elsevier and the journal *Science*): conceived and designed the study, acquired the funding, collected the data, contributed data or analysis tools, performed the analysis, wrote the paper, and so forth. Disclosures should include any financial interests and outside activities or interests that relate to the research presented. When there are no conflicts, a simple statement often suffices, such as:

> The authors declare no conflicts of interest.

3.1.7 Transparency, Rigor, and Reproducibility

Since the 1st Edition of this book appeared in 2009, there have been many changes in scientific reporting and hence publishing, particularly related to "transparency." Given the importance of these changes, please find further discussion in Chap. 4 (Proposals and Grant Applications) and Chap. 7 (Managing Data). Importantly, the National Institutes of Health (NIH) invited editors from the *Nature* Publishing Group and the journal *Science* to a workshop in June of 2014 to identify and promote opportunities "to enhance rigor and further support research that is reproducible, robust, and transparent," see

www.nih.gov/research-training/rigor-reproducibility/principles-guidelines-reporting-preclinical-research.

An important outcome of this workshop was broad adoption by journals of life science and medicine of new requirements for publication that are designed to advance three goals: "to promote rigorous statistical analyses, to encourage wider availability of data and materials, and to increase transparency in reporting." In particular, such transparency was designed to achieve six goals: "to encourage the use of community-based standards, to require details on the number of biological or technical replicates that define data sets, to document whether samples were randomized, to require information on possible blinding of investigators to experimental groups, to provide details on how sample sizes were determined, and to stipulate any inclusion or exclusion criteria." For additional details, the reader is referred to a concise summary published early in 2022 from the perspective of American Heart Association journals (Lu & Daugherty, 2022). In many cases, journals have established related checklists that authors must complete as part of the submission process.

Various guidelines have also emerged from these discussions, including TOP (Transparency and Openness Promotion) and ARRIVE (Animal Research: Reporting of In Vivo Experiments). TOP suggests eight different standards to increase rigor and reproducibility, each with multiple levels of possible stringency. For example, TOP proposes three levels of transparency related to data availability: "to disclose whether the data associated with a publication are available and, if so, how to access them (Level 1), to deposit the data in a valid and accessible repository (Level 2), to deposit data in a repository only after a third party has validated its reproducibility (Level 3)." For further information on TOP, see www.cos.io (Center for Open Science). Related to these guidelines, some journals are now requiring that papers be posted as non-reviewed preprints prior to submission for peer review and possible publication in an archival journal. One such preprint repository is bioRχiv (www.bioRxiv.org), which was established in November 2013 and is maintained by Cold Spring Harbor. The rationale is that open access preprints can be read and commented on by an audience much broader than the typical 3–5 reviewers that serve in an advisory capacity to a journal editor, thus encouraging broader assessments from a more diverse audience.

The ARRIVE guidelines similarly seek to increase rigor and reproducibility, but in this case specifically in studies using animals. That is, in addition to merely stating that animal studies reported in a manuscript were approved by the local institutional review committee, authors are now encouraged to address ten key aspects of the research: "study design, sample size, inclusion and exclusion criteria, randomization, blinding, outcome measures, statistical methods, experimental animals, experimental procedures, and results." Each of these ten aspects of research entail specific recommendations. For example, for experimental procedures, it is recommended that the Methods include (again, for animal research) detailed descriptions of what was done, how was it done, when and how often was it done, where was it done, and why was it done. For further information on ARRIVE, see https://arriveguidelines.org.

3.1.8 Appendices and Supplements

Most scientific papers today do not contain appendices. Nevertheless, when used well, appendices can serve a very important role. In general, appendices contain important information that either does not fit well within the flow of the body of the paper or is simply best stated separately for those few readers who will be interested in such details. A good example of an Appendix would be the step-by-step derivation of key equations, the final result of which should be discussed in the body of the paper. In this way, authors fulfill their responsibility of providing methods that are sufficiently detailed to enable the reader to reproduce a result while not distracting the reader from key points presented in Methods. Similarly, detailed recipes for molecular or cellular assays may fit well in an Appendix.

More common since the advent of on-line publishing and the availability of extensive cloud-based storage is the use of supplements, often referred to as Supplementary Materials or Supplemental Information. These supplements typically contain figures and tables that provide important information that is nevertheless not critical for the flow of the main body of the paper. Indeed, given the increased call for transparency and data availability, these supplements can fulfill many of the newer requirements imposed by journals while, most importantly, providing rich data sets that can be re-used by other investigators. In some cases, Supplementary Materials can also include much more detailed descriptions of methods used and additional references. It is critical for the author to know, however, that Supplementary Materials are typically not checked or copy-edited by staff at the journal. Rather, these materials are submitted in final form as PDF files and simply linked to the published paper on the journal web-site—they thus demand a high level of proof reading prior to submission.

3.1.9   References

It is interesting that we are often taught "proper methods of citation" in courses and books despite the many different journals and publishers requiring very different citation formats. In some cases, references must be arranged according to the order of appearance within the work and numbered sequentially beginning at 1; in other cases, references must be arranged alphabetically by the last name of the first author, then numbered beginning at 1; in yet other cases, references must be arranged alphabetically and not numbered. This basic scheme dictates how to cite any reference within the text, by number or by author. Similarly, the format for the references that details the authors, year of publication, title, volume, and inclusive pages also varies from journal to journal. The best advice, therefore, is to follow the specific *Instructions for Authors* on the journal's website. As specific examples, however, consider multiple ways to cite the same article within the text:

> Watson and Crick (1953) proposed the double helix …
> Watson and Crick [20] proposed the double helix …
> Watson and Crick[20] proposed the double helix …

or similarly,

> The double-helix structure of DNA was proposed in 1953 (Watson and Crick, 1953).
> The double-helix structure of DNA was proposed in 1953 [20].
> The double-helix structure of DNA was proposed in 1953[20].

As seen, the third approach in both cases results in some savings with regard to printing, which is important to some publishers given that most journal articles cite ~35 papers and most review articles cite over 100 papers. When these simple savings are multiplied 30-fold or more, one can appreciate the potential savings in page costs for print journals. Nevertheless, numerical citations have the disadvantage that the reader must constantly refer to the reference list to determine who was responsible for the cited finding. Informed readers often know who has done what in a field, which is to say who has produced reliable or important findings. Citation by author names (e.g., Smith et al., 2021) thus has the advantage of increasing the flow of the paper. Yet, one must follow the format prescribed by the journals and publishers. Usually the only case wherein one can pick a format is while writing proposals, which is discussed in Chap. 4.

Citation is similar to that discussed above when there is but a single author. For example, we might find the following: Einstein (1905) proposed..., Einstein [20] proposed..., or Einstein[20] proposed..., and similarly we might find...the special theory of relativity (Einstein, 1905),...the special theory of relativity [20], or...the special theory of relativity[20]. In the case of three or more authors, however, the format differs slightly.

Recall from Chap. 2 that "and others" is abbreviated in the Latin as "et al." (which may or may not be italicized, depending on the journal). Hence, we might find the following: Smith et al. (2021) proposed... or... (Smith et al., 2021). Whereas some journals use (Smith et al. 2021), that is, they omit the comma after the et al., it is a mistake to add a comma after the first author's last name (i.e., Smith, et al., 2021 is not an accepted format). Again, the best advice is to refer the specific Instructions for Authors for the journal of interest.

Finally, note that the citation within the reference list can also appear in various forms. For example, we can cite the same paper as:

Watson JD, Crick FHC (1953) Molecular structure of nucleic acids: A structure for deoxyribose nucleic acid. Nature 171: 737–738.
Watson JD, Crick FHC. Molecular structure of nucleic acids: A structure for deoxyribose nucleic acid. *Nature*. 1953;171:737–738.
Watson, J.D., and Crick, F.H.C., 1953, "Molecular Structure of Nucleic Acids: A Structure for Deoxyribose Nucleic Acid," Nature, **171** pp. 737–738.

Other formats exist, which is why one must consult the *Instructions for Authors* for each journal.

A final, and important, reminder is to cite only those papers that you have actually read. Perhaps surprisingly, many investigators will cite papers that someone else has cited simply because this appears to be easier. Such a practice can be dangerous, however. In science and engineering, one should always check and double-check everything, including interpretations of other work used in citations.

3.1.10 Figures and Tables

It has been said that a picture is worth a thousand words. Actually, a well-prepared and appropriately selected picture (e.g., figure or image) can be worth a thousand words if done well. As an example, consider the following figure, a standard x–y scatter plot, which is the most common type of figure found in a technical paper. Although this example contains the basic ingredients of an effective figure (e.g., clear data points and labeled axes, with the unit of measurement denoted parenthetically on the x-axis), we can improve it considerably with little effort.

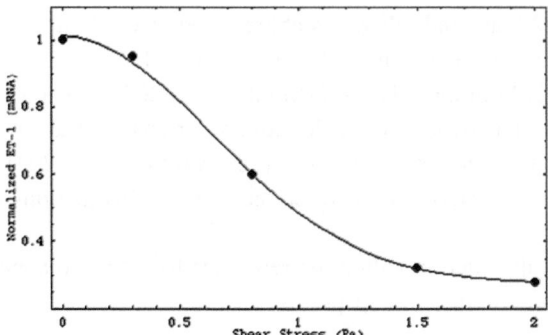

Compare the following version of this figure (reprinted with permission, CISM) to the previous one. It is easy to see that a reduced number of tick marks along each axis as well as larger numbers and lettering improve the readability considerably. Indeed, one of the most important considerations is that typesetters will reduce the size of many submitted figures before placement within the final version of the paper. This is particularly important when placing a figure within a single column in a dual-column layout, which most technical journals use.

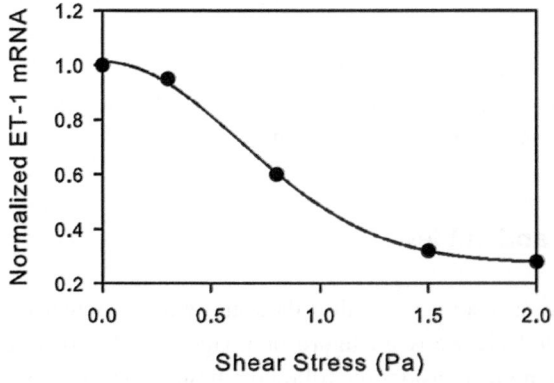

Albeit a small point, note that the solid curve in these figures represents a best-fit to data obtained using a formal regression method. Whereas solid lines are appropriate for showing such theoretical or model fits, it is best to use lightly dotted or dashed lines when the goal is to connect the data points for emphasis. It should be remembered, however, that simply connecting the dots may be misleading if data sampling missed key points.

Also in the spirit of transparency, most journals now require data presented in bar plots to include all of the individual data points rather than simply the mean and standard deviation or standard error of the mean. A discussion of what data to include, or exclude, is found in Chap. 7. Finally, realizing that many readers go to Results after having only read the title and perhaps the Abstract, it is important to write complete legends so that the reader can understand the importance of the figure easily.

3.1.11 The Archival Paper Summarized

The original journal article should be both written and illustrated with care; regardless of the order of presentation it should address the following primary questions:

Introduction: What was done and why?
Methods: How was the work accomplished?
Results: What was found?
Discussion: Why are the results important, how do they compare to or with previous work, and what remains to be investigated?

Seeing one's name in print for the first time on an archival paper can bring a sense of excitement and pride. Seeing one's name on a paper that contains errors or fundamental flaws can bring a sense of regret. There is, therefore, a need to give such work our most careful attention from start to finish.

Exercise 3.1

Most journals limit the number of words or pages allowed for papers within particular categories. For example, prior to on-line publishing, it was common for an original article to be limited to 6000 words inclusive. If an abstract typically is ~250 words, each one-column table and a single panel figure is equivalent to ~250 words, and a standard full-citation reference is typically equivalent to 20–30 words (some up to 40 words), estimate the number of words for an abstract, 6 figures, 1 table, and 35 references, which is typical for a standard original paper. Next, calculate the number of words available and estimate reasonable lengths for remaining sections: Introduction, Methods, Results, and Discussion.

Exercise 3.2

Research and summarize the TOP (Transparency and Openness Promotion) guidelines and discuss in a three-page report (double-spaced, one-inch margins, 12-point font) how science and engineering may benefit if journals uniformly impose such requirements.

Exercise 3.3

Research and summarize the ARRIVE (Animal Research: Reporting of in vivo Experiments) guidelines and discuss in a three-page report how science and biomedical engineering may benefit if journals uniformly impose such requirements.

Exercise 3.4

Find three published papers of interest that include Supplementary Materials and discuss in a three-page report any differences in the quality or utility of the figures that are found in the main body of the paper versus those in the supplement, with the goal of developing a personal strategy of augmenting primary results with supplemental information.

3.2   Publishing an Archival Journal Paper

3.2.1   Origin

According to Boorstin (1983, pp. 390–394),

> The printed scientific 'paper' or 'article', which was simply a later version of the letter, would be the typical format in which modern science was accumulated and communicated. ... The letter was an ideal vehicle for the increasing numbers of men dispersed over Europe who no longer expected to storm the citadel of truth, but hoped to advance knowledge piece by piece. ... A letter had obvious advantages over a book. While works of science were often large tomes easy to stop for censorship, the novel observations in a letter could slip unnoticed or be delivered with the 'ordinary post'.

In contrast to early European investigators such as Galileo (1564–1642), few modern investigators working within democratic societies need be concerned about potential censorship of their work. Rather, the primary concern today is to ensure that a paper receives broad distribution to the intended audience. Toward this end, electronic publishing and the Internet have revolutionized the availability of scientific papers, yet the methods of composition, presentation, and submission have not changed.

3.2.2 Composition and Authorship

It is difficult, if not impossible, to write by committee. Indeed, one of the most important documents in American history, the Declaration of Independence, was assigned to a committee of five for composition but was drafted in seclusion by a single author, Thomas Jefferson. As it should have done, the committee evaluated and revised the final draft penned by Jefferson before forwarding the final version for consideration by the Continental Congress of 1776.

Although there has been a significant increase in the number of authors on scientific papers, particularly in biomedical science and engineering, the primary responsibility of writing a paper must similarly fall to one author or in some cases two authors (e.g., the first and senior authors). As noted above, however, the best way to ensure that the first draft represents the ideas and expectations of all authors is to meet together to define the initial outline and to discuss what findings to report in Results. We address the issue of joint authorship further in Chap. 6, hence we merely note here that it is essential that all authors agree on the contents and presentation of a paper before submission for consideration of publication.

3.2.3 Submission and Review

As noted earlier, the essential first step when preparing a paper for consideration for publication is to read *Instructions for Authors* for the intended journal. Only in this way will one be able to fulfill the requirements of each journal. In general, however, the two primary items needed for submission are a cover letter to the editor and the complete paper, including the cover page, body of the paper, tables, and figures.

The cover page serves to communicate to the editor and publisher important pieces of information: the title (and thus subject) of the work, those who performed the work (i.e., an author list in a specific order) and their professional affiliations, keywords that further classify the area of study, and finally the full address of the corresponding author. The title is extremely important for it will determine to a large extent who reads the paper. A good title captures the essence of a paper without being overly long. Indeed, a general rule of thumb is that the title should not exceed 120 characters. In the past, it was also suggested that a title should not contain verbs. Consider, for example, well-known titles from two of the most important papers from the 20th century:

"On the Electrodynamics of Moving Bodies"
"Molecular Structure for Nucleic Acids: A Structure for Deoxyribose Nucleic Acid"

The first example is from A. Einstein's famous paper of 1905 that introduced his special theory of relativity; its title is brief and general. The second example is from J. Watson and

F. Crick's famous paper of 1953 that introduced their concept of the double-helix structure of DNA; its title is also brief and general, but the subtitle provides more specific insight into the focus of the paper. Interestingly, neither of these seminal papers are among the top 100 cited papers. At the time of this writing, the top cited paper (>300,000 citations) was the 1951 paper "Protein Measurement with the Folin Phenol Reagent," which again is brief and general. The *American Medical Association Manual of Style* (1998) recommends further that titles should not contain phrases such as "The Role of..." or "The Effects of..." or "The Treatment of..." and so forth. In contrast to earlier suggestions that titles should not contain verbs, titles of more recent papers in biology often use verbs for impact. Although there is no need to be dogmatic when crafting a title, simple guidelines are useful to consider nonetheless.

Select keywords that are distinct from words used in the title and based on general, but not generic, aspects of the paper to ensure broader recognition. It is both much easier and more important today to identify appropriate keywords. One can log onto a standard on-line search engine, such as PubMed, and compare results for different keyword searches to identify those that highlight papers most closely related to your work; such words would be good candidates for your keywords. Because most investigators now search for technical papers using such search engines, we cannot overemphasize the importance of appropriate keywords. That is, writing well is not enough—if the work does not reach the intended audience, it will not have an impact. Significant attention must be given to the title, keywords, and, as noted below, the Abstract.

Most journals require a cover letter. This letter serves, in part, as the official "intent to submit" the paper and thereby to agree to all policies and procedures of review adopted by the selected journal. Among the many points that can be addressed in this letter, it is customary to confirm when the work is original, that all authors contributed to the work and agree to its submission, and that the paper is not simultaneously under consideration for publication elsewhere. That is, just as with submissions of grant applications to most agencies, there is a standing agreement among scientists and engineers that it is improper to submit the same work for simultaneous evaluation by two or more journals because of the significant effort required of others to provide proper and timely reviews. One may also wish to note why the work will be of interest to the readership of the journal or to identify potential reviewers who either should or should not be selected and why. Like the paper itself, however, the cover letter to the editor should be concise. Here, we provide a simple example; letters will vary depending on individual circumstances.

Date

Dr. J. Smith
Editor, *Journal Name*
Address

Dear Dr. Smith:

Enclosed please find the manuscript, "Title," which is submitted for consideration for publication in *Journal Name*. This paper represents original research that has not been, nor will it be, considered for publication elsewhere until a decision is reached by you or your staff. All authors contributed to the work and its preparation and agree to its submission.

With very best wishes,

Sincerely,

Name

Title

If applicable, the letter to the editor should also contain statements that all research involving human subjects or animals conformed to accepted standards and was approved by the appropriate institutional committee. Similarly, if applicable, this letter can communicate that permissions have been obtained from the appropriate parties to republish previously published work. If there are or are not conflicts, or potential conflicts, of interest, this can be stated as well. More recently, many papers are deposited on-line as open access preprints prior to submission of a paper to an archival journal (recall Sect. 3.1.7). If this is the case, it is appropriate to note this in the cover letter. Finally, the cover letter should be on institutional letterhead; it is an official document.

An editor or associate editor will usually solicit three or more experts to provide a recommendation on the potential suitability of a paper submitted for publication. These reviewers are asked to provide objective assessments and thus to decline to review a paper if there is either a real or a perceived conflict of interest. In the past, reviews were almost always anonymous. This practice has changed some in recent years and further changes may continue in coming years, again in part due to the desire for full transparency. The period allowed for review varies considerably across different journals, with some biological and clinical journals allowing only 10–14 days and some mathematical journals allowing 3 months for review. Differences also exist across journals with regard to the possible categories of recommendations available to the reviewer, but general categories are:

Accept (i.e., accept as submitted)
Accept pending minor revision (not requiring re-review)
Major revision (with required re-review for further consideration)
Reject (i.e., not suitable for publication)
Inappropriate for this journal.

In cases where the topic, type, or length of a paper is deemed to be inappropriate for a journal, an editor can communicate this to the author(s) without a formal review, even though reviewers are also allowed to make such a recommendation independently. A recommendation to reject a paper may be rendered for any of a number of reasons: the paper does not contain original or novel findings, it contains serious flaws in design or analysis of the data, it does not address a relevant problem, the results contradict well-accepted findings without addressing the associated reasons, there is insufficient new information (i.e., it is only an incremental advance at best), and so forth. In the case of a recommendation to reject, the editor should ask the reviewer(s) to state this case diplomatically, although this does not always happen. It is common for a journal to reject 50–70% of all submissions, with highly selective journals rejecting 80% or more.

The recommendation to request major revisions generally implies that there is a need for additional experiments, analysis of data, or computations. Such a recommendation can also reflect the need to correct a major, but not fatal, flaw, to reduce the length of the paper if it is overly long, or to improve significantly the presentation, including improved figures, tables, and writing.

By contrast, minor revisions (not requiring further review) can reflect a need to expand Methods or Discussion or conversely to eliminate information that is available in other publications. There may also be a need to add some key references or reduce the overall number of references. Minor improvements in writing or the need to eliminate results that are duplicated in tables and figures may also lead to a recommendation of the need for minor revisions.

Finally, albeit uncommon, the recommendation to accept (as is) suggests that the study is important, novel, and well presented. All authors should strive to submit such work, but should be prepared to revise a paper as needed.

As one might expect, unanimous recommendations by two or more reviewers are uncommon, hence the editor or associate editor usually must make a decision based on the information received. For example, if two reviewers recommend major revisions and a third reviewer recommends rejection, the editor is justified in recommending either major revision or rejection. In cases where the editor rejects a paper, the authors can try to rebut the reviews and request either the privilege to submit a revised paper or that additional reviewers are asked to review the paper further. In most cases, however, editors tend to stand by the initial, carefully weighed decision, and it is best to consider other options. For example, some authors will simply resubmit the same paper to another journal for consideration for publication; in such cases, they are usually not required to reveal that the paper has been rejected previously, which enables the second assessment to be unbiased. Authors must realize, however, that reviewers are often picked carefully for their expertise, and it is possible that different editors from different journals will select the same reviewers. For this reason, and simply because one should always use any opportunity to improve a paper based on any feedback obtained, it is best to revise a paper that has been rejected before submitting it to another journal.

Finally, it is useful to know how editors select reviewers and what instructions are given to the reviewers. Ideally, the editor, associate editors, or editorial consultants are familiar enough with the topic of the submitted paper that they know the experts personally or at least know of them. In cases wherein such experts either decline to review a paper (because of conflicts or simply because they are not able to provide a timely review) or cannot be identified easily, editors will often peruse the references cited in the paper. In other words, frequently cited authors are good potential reviewers because their work is related closely to that which was submitted. Alternatively, editors may also use computerized search engines to identify potential reviewers based on the publication of similar work in reputable journals.

> **Exercise 3.5**
> Interview someone who serves on an editorial board of a technical journal and ask how reviewers are selected, what are good reasons for excluding reviewers in particular cases, and what is done when different reviewers have diametrically opposing views. Write a one-page summary of the results of the interview.

> **Exercise 3.6**
> A few journals allow authors to submit a paper for consideration directly to a member of the editorial board or the sponsoring society. In these cases, that member can assume the sole responsibility for review and may then "communicate" the paper to the editor for publication. Identify two different journals that allow such a procedure and write a two-page summary discussing the history of this approach and the associated advantages and disadvantages.

As noted above (Sect. 3.1.7), in the interest of increasing rigor and reproducibility, some journals now ask that papers be submitted to non-reviewed, open access preprint servers with the expectation that a broader community can provide additional assessments beyond that provided by specially selected reviewers.

> **Exercise 3.7**
> Research and discuss the potential advantages and disadvantages of open access peer review and submit a three-page summary.

3.2.4 Revision

As noted earlier, only a small percentage of technical papers are accepted upon the first submission, hence authors should expect to revise a submitted paper. Indeed, in most cases, revision along the lines suggested by the reviewers improves the paper significantly, thus revision should be seen as an opportunity not a failure. Nevertheless, it is human nature to be disappointed or, in some cases, upset by a negative review. Toward this end, we recommend two things. First, read the review carefully but do not take any action until at least a few days later. In other words, neither a rash response to the editor nor a hasty attempt to revise a paper is likely to be productive. Second, avoid both of the most natural responses to a negative review—to adopt all of the reviewer's recommendations because "they must be the expert" or to ignore all of the comments because you "know better." Rather, it is best to take comments and concerns by reviewers at face value. For example, if a reviewer states that a particular section is hard to understand even though you think it is clear, chances are that at least some other readers will also find the section hard to understand. The best response in this case is to take advantage of the opportunity to improve its clarity and to ask a colleague to assess the changes. In other words, because it is your name on the paper, take every opportunity to make the paper the best it can be.

When a journal allows or requests a revision, the author(s) usually must submit the revision within a certain period (often 1–3 months, but highly variable) and provide evidence of the revisions. In some cases, the author(s) can meet this requirement simply by summarizing the revisions on a separate page or in a letter to the editor. In most cases, the journal requires both detailed responses to each of the reviewer's concerns and marks within the submitted manuscript that identify the revisions. The latter requirement is now met easily using features such as "track changes" or simply by using a different color font for any revisions. Importantly, some journals are now publishing both the reviews and the responses to the reviews as supplemental materials. It has always been important to write thoughtful, clear, respectful responses; whether the responses will be published or not, we recommend the same level of care in crafting responses.

Once the authors decide to revise a paper, they should ask how to do this most efficiently. For major revisions, it is best to identity the requisite experiments, calculations, analyses, and so forth and to generate the additional results. Next, one can follow the same approach used in writing an initial draft—lay out all results, new and old, on a table (or computer screen) and determine which to include and in which order. Once done, first revise the Results, Methods, and Discussion, then the Introduction and Abstract accordingly, and document the revisions appropriately. For minor revisions, it can be efficient to begin by writing a "Response to Reviewers" or "Summary of Revisions." After knowing how one wants to address the concerns, the paper can then be revised accordingly.

Although policies differ among journals, it is uncommon to allow authors to submit more than one revision because of the time invested by both the editors and the reviewers. In other words, one should work very hard to satisfy reviewer's concerns and to make a revised manuscript acceptable. Indeed, although the original version should be submitted only after careful construction and correction, a revision should be considered final and thus written such that you are proud of the final submission.

Exercise 3.8
Read the article "Structural Outline of an Archival Paper for the *Journal of Biomechanics*" [Brand RA, Huiskes R (2001) *J Biomech* 34: 1371–1374]. Construct a

two-column table to record hints therein that reinforce or contradict that which was presented in this Sect. 3.2. Provide a summary, not to exceed one page, that articulates your preferences in cases where there is disagreement.

3.2.5 Typesetting and Final Approval

Years ago, a typewritten copy of a manuscript was copyedited, then typeset from scratch. Today, journals require electronic versions of accepted papers, which are then copyedited and formatted by the publisher prior to publication. Briefly, copyediting is an important step wherein a manuscript is checked carefully by the editorial staff of the journal, or a third-party associate, for format, style, spelling, complete and consistent citations, and so forth. In some but not all cases, the copyedited version is sent to the authors along with either the html file (for on-line corrections) or a galley proof (often PDF file) to show the changes that were made.

A so-called galley proof is the final draft of the paper formatted as it will appear in print or on-line. The authors are asked to check this final document carefully to ensure accuracy, yet it is expected that only minor changes or corrections will be made at this stage of publication. In the case of major changes due to author errors, the publisher may charge extra to make the requested changes. For this reason, authors should ensure that the final version of a manuscript is correct as submitted. Although the advent of electronic publishing decreased tremendously the number of necessary corrections, authors should be diligent to check the html file or galley proof carefully, including the layout of tables, figures, and equations. If errors are discovered after final approval and publication of the paper, the authors' only recourse is to publish an erratum.

3.2.6 Copyright, Permissions, and Publication Charges

Copyright is a legal procedure that grants exclusive rights to the production, publication, sale, or distribution of a work by the owner of the copyright. Upon acceptance of a paper for publication, the publisher will usually request that the authors transfer copyright. Although copyright agreements tend to be standard for the publication of scientific papers, one should read such agreements carefully before signing. If there are questions regarding anything within the agreement, one should either consult a more experienced author or an appropriate institutional official, often after contacting the copyrights division of the publisher for further clarification.

Transfer of copyright requires that the material to be transferred is original or, when appropriate, that special permissions have been obtained to republish any previously published material. In the latter case, the most common situation is the desire to republish a figure or image from another work, often for the purposes of a Review rather than an Original Article. One may obtain permissions to do so either by using Creative Commons (www.creativecommons.org) or by contacting the copyrights division of the publisher that holds the copyright and requesting permission to republish the work in a specific way. In most cases, a publisher will grant such permission provided that a simple statement is associated with the republished material [e.g., "From Smith (2021), with permission from *Publisher Name.*"]. In some cases, however, such permission is granted only after receipt of a fee, which could be hundreds of dollars for a single figure. In cases of financial constraints, it is best to identify potential fees early. Although one only needs to contact the holder of the copyright, which is usually the publisher not the original author(s), requesting permission from the author(s), if they can be located, is a good gesture.

In many cases, one step remains before the publication of your paper—payment of fees. Such fees can now be divided into two types, those associated with print versions (i.e., the subscription model) and those associated with Open Access. The former include both general "page charges" and specific fees to cover the publication of color figures in print versions. In many cases, standard page charges depend on the length of the paper and are designed to offset portions of the cost of publication of a paper that appears in print in a journal. Such page charges are mandatory in some cases but voluntary in others. The rationale behind voluntary page charges is that many agencies that financially support research also desire that the associated findings be published, and consequently, they provide researchers with funds to support publication. Payment of page charges often entitles the author(s) to free hardcopy reprints or PDF versions of the paper. Independent of page charges, some journals assess mandatory fees for the publication of color figures in print (but not electronic) versions. Because fees for color figures can be hundreds to thousands of dollars, it is prudent to minimize the use of color figures or images to those that are best understood in color. With the continued growth of online publishing, however, one can also consider using color figures in the online version that are equally clear if printed in black and white by the reader or by the publisher of the print version. Regardless, it is good to determine before submission if such fees will be charged. Like all submitted figures and images, color versions must have the required resolution and must be submitted in the proper file format; remember, too, that a certain percentage of the readers will be color-blind, hence choose color schemes appropriately. Again, the reader is referred to the *Instructions for Authors* for each journal for requirements vary with journal.

So-called "Open Access" to scientific papers dates back many decades, but this change in publishing received strong support primarily over the past two decades (Budapest Initiative in 2002, Bethesda Statement of 2003, and Berlin Declaration in 2003). Indeed, Open Access publishing continues to grow significantly, and all potential authors should consider this option before selecting a journal to which to submit a paper. In most cases, fees for Open Access publication are well over $1000 US per paper, although some publishers now offer institutional rates such that any author associated with that institution has the publication fees waived. Importantly, some funding agencies now require that associated papers be published open access (e.g., in Europe). Others, such as the NIH, currently require that the published paper be made available following a particular embargo period, often 12 months. Papers published under this model receive a unique PMCID number (e.g., PMC1000000), which can be found on the PubMed website.

Exercise 3.9
Research the history of on-line publishing of scientific and engineering papers and write a three-page report summarizing advantages and disadvantages.

Exercise 3.10
Research the history of open access publishing of technical papers and write a three-page report summarizing advantages and disadvantages.

3.3 Thesis or Dissertation

Most universities require the completion of a thesis as part of the requirements for a master of science (M.S.) degree or a dissertation for the completion of a doctor of philosophy (Ph.D.) or a doctor of science (D.Sc.) degree. It is common for an M.S. thesis to range from 50 to 150 pages and for a Ph.D. or D.Sc. dissertation to range from 100 to 300 pages (each double-spaced with ample margins). Although it may seem a formidable task to write such a long document, it is actually not difficult if one formulates a good outline and simply writes chapter by chapter. One should obtain specific guidelines for formatting such documents from the local Office of Graduate Studies, however, because requirements differ from institution to institution. Here we note briefly the two most common styles for organizing a thesis or dissertation.

First, a thesis or dissertation can be organized along traditional lines and thus consist of the following:

Abstract
Chapter 1: Introduction
Chapter 2: Background
Chapter 3: Methods

Chapter 4: Results
Chapter 5: Discussion
Chapter 6: Conclusions and Recommendations
References
Appendices

Note the strong similarity between this outline and that for the archival journal paper, with two notable exceptions. To encourage students to understand the literature well, most institutions require a separate chapter, titled Background, which highlights past work on the topic of graduate study and identifies areas of need for further study. Because a dissertation must represent original work, review of the background literature is a particularly important part of the doctoral student's early work. To encourage students to recognize limitations of their own work and thereby to identify areas for further study, most theses and dissertations end with a chapter titled "Conclusions and Recommendations." A simple way of thinking about recommendations is to ask the question, "What should the next graduate student in the laboratory do to push the current work forward?"

In contrast to the traditional outline, it is becoming increasingly common to organize theses and particularly dissertations around individual papers that are based on the student's work and either have been or will be submitted for publication in archival journals. For example, consider the following outline:

Abstract
Chapter 1: Introduction
Chapter 2: Paper 1
Chapter 3: Paper 2
Chapter 4: Paper 3
Chapter 5: Conclusions
Appendices

Packaging the multiple papers (often one for a thesis and three to seven for a dissertation) between general introductory and concluding chapters allows a student to focus on writing individual journal papers. In this case, each chapter contains its own Introduction, Methods, Results, Discussion, and References. It is clear that the individual papers should be written first, as described above, and the Introduction and Conclusions written last. In the traditional case, it is common to write the Background first, then Methods and Results, and finally Discussion, Introduction, and Conclusions and Recommendations. If one adopts this second style of organization and papers are published before submitting the final dissertation, it is important to consider issues of copyright; the best advice in this case is to check with your local Office of Graduate Studies.

Exercise 3.11
For graduate students or students who are interested in pursuing a graduate degree in science or engineering, find and read carefully the instructions for a thesis or dissertation at your institution. Write a one-page bulleted summary of the key stylistic requirements.

Exercise 3.12
Given the increasing number of authors on many scientific and engineering papers, write a three-page discussion of challenges that emerge when such work is part of a thesis or dissertation when it comes time to write, submit, and defend the final document.

3.4 Technical Reports

Whether a technical report will be published or not, it is a formal document and thus requires careful attention. In contrast with the aforementioned types of technical documents, however, the technical report does not demand a particular outline. Indeed, such reports can range from a one-page summary to a thousand-page document. The best advice, therefore, is to discuss in detail the expectations before beginning the outline, then use, as appropriate, the aforementioned guidelines for writing an archival journal paper. In general, material contained within a technical report is neither copyrighted nor published and thus is often available for subsequent submission for publication. It is important, however, to confirm whether there are any restrictions on possible publication, particularly when the report is submitted to a company.

References

Boorstin, D. J. (1983). *The Discoverers.* Vintage Books.

Lu, H. S., & Daugherty, A. (2022). Key factors for improving rigor and reproducibility: Guidelines, peer reviews, and journal technical reviews. *Frontiers in Cardiovascular Medicine, 9*, 856102. https://doi.org/10.3389/fcvm.2022.856102

Proposals and Grant Applications

4

4.1 Introduction

Whether it be a proposal to undertake a project as part of senior design by an undergraduate student, to pursue a specific topic for a thesis or dissertation in partial fulfillment of the requirements for a graduate degree, to define a new direction for research and development (R&D) by a division within a company, or to support basic or translational research by a university professor, the technical proposal is fundamental to securing the approvals and resources needed to advance science and engineering. Notwithstanding the many different types of proposals that are possible, most are similar in basic structure and preparation. Hence, for illustrative purposes, we focus on the National Institutes of Health (NIH) individual investigator grant application known as the R01, the primary mechanism for funding health-related research in the United States.

4.2 Types of Grants

The R01 mechanism supports two general types of grant applications: investigator-initiated and those in response to a request for proposals (RFP) or program announcement with special review or receipt information (PAR). By investigator-initiated, we mean a scientist or engineer identifies a fundamental question or important problem, conceives of an approach to address this issue, and takes the initiative to request the funding needed to complete the project. By contrast, a RFP or PAR is a "call for proposals" that focuses attention on a particular need or area of investigation that a working group, committee, or administrator has deemed to be important. In the latter case, the scientist or engineer must first learn of the opportunity, then respond according to the stated instructions. Indeed,

J. D. Humphrey and J. W. Holmes, *Style and Ethics of Communication in Science and Engineering*, Synthesis Lectures on Engineering, Science, and Technology, https://doi.org/10.1007/978-3-031-39125-5_4

one of the most important aspects of a successful application submitted in response to a call for proposals is that it is responsive to the announcement.

Exercise 4.1
Go to the NIH website, www.nih.gov, and search for current PARs. Pick a PAR of interest and read the instructions carefully. Write a three-page summary of the PAR that could serve as a sufficient overview of the motivation, scope, and requirements for submission. Use an 11 point Arial font, 1.0 line spacing, and 0.5-inch margins, which are accepted by the NIH.

Exercise 4.2
The National Science Foundation (NSF) funds basic research and training in the sciences and engineering in the United States. Go to the NSF website, www.nsf.gov, and search for current funding initiatives. Pick an initiative of interest and read the instructions carefully. Write a three-page summary that would be sufficient as an overview of the motivation, scope, and requirements for submission.

There are three general classes of motivations for any proposal in science and engineering: hypothesis-driven, curiosity-driven, and technology-driven. No one motivation is more important or more scholarly than another, they are simply different.

Hypothesis-driven—The NIH defines a hypothesis as "an educated prediction about the outcome of your study." Under some programs, the omission of a hypothesis is viewed as a major oversight, one that can result in the reviewers suggesting that the proposal is merely a "fishing expedition," that is, a project without clear direction.
Curiosity-driven—We have all heard the saying that when asked why they climbed a mountain, climbers may simply state "Because it is there." Curiosity-driven research reflects a desire to find answers to new and intriguing questions. Curiosity has been, and will remain, a primary driver of scientific inquiry and engineering development.
Technology-driven—In our increasingly technology-based society, there are many cases wherein the ability to design and build a new instrument or system is motivation enough to pursue such research. Indeed, in many cases there are excellent opportunities to modify previous designs to address new applications.

Regardless of the motivation—hypothesis, curiosity, or technical need—the approach to apply for and secure funding for research is similar.

4.3   The Review Process

Proposals undergo a two-step review process at the NIH. First, proposals are evaluated for technical merit and feasibility by a panel of discipline-specific experts that is called a Study Section, which is managed by a government appointed Scientific Review Officer (SRO). Second, proposals are evaluated administratively for funding potential by a Council that seeks to balance the desire to fund the best science with accomplishing the mission of the NIH—namely, "to seek fundamental knowledge about the nature and behavior of living systems and the application of that knowledge to enhance health, lengthen life, and reduce illness and disability."

Just as we should identify the intended audience before writing a journal article or giving an oral presentation, we should also understand the audience that will review a particular proposal. In contrast to many funding agencies, the NIH publishes the names of those constituting most Study Sections, which enables applicants to know their potential audience. In this case, it is prudent to read recent publications by members of relevant Study Sections to get a feel for their scientific interests and basic perspectives. NIH allows applicants to suggest particular Study Sections via a "PHS Assignment Request Form," hence it is important to make informed recommendations. Although each member of the committee will be asked to score all grant applications for which they do not have a conflict of interest, only three to five individuals generally read each application in detail. These individuals are referred to as the primary reviewer, secondary reviewer, and discussant(s); they are selected based on the closeness of their technical expertise to that of the proposal, as revealed primarily by its title (200 characters maximum) and Project Summary (30 lines maximum). Recalling that we use the NIH R01 herein mainly to illustrate issues that are important in preparing a proposal, graduate students would similarly be well advised to know the composition of their committee and to read recent papers by those individuals to anticipate what types of questions might arise.

Whether or not one knows the composition of a review panel, among the most important things to know are the criteria that the panel will use to evaluate the proposal. Again, we use the NIH as our illustrative example, but the need to be familiar with such criteria would similarly apply to other funding agencies as well as to a master's or doctoral research proposal that a faculty committee would evaluate during a proposal defense. Currently at the NIH, the evaluation criteria include[1]:

Significance—Does the study address an important problem?
Innovation—Does the project employ novel concepts, approaches, or methods?
Approach—Are the design and methods appropriate to address the aims?
Investigators—Are the investigators appropriately trained?

---

[1] Portions of this discussion were based on a web-based document, *Proposal Writing: The Business of Science*, by Wendy Sanders, then at the NIH.

Environment—Will the scientific environment contribute to the success?
Overall Evaluation—Will the study advance health care or medical science?

Additional considerations of importance during a review include whether proposed human or animal research protocols follow current guidelines. Before beginning any such study, a local institutional committee (e.g., Institutional Review Board, or IRB, for Human Subjects and Institutional Animal Care and Use Committee, or IACUC) must evaluate and approve such protocols, but this need not be completed before submission of the proposal. The applicant can merely complete, as appropriate, the Human Subjects and Vertebrate Animals sections and list the approvals as pending. The reviewers will also assess whether the proposed budget is justified. We focus here, however, on the scientific aspects of the review enumerated above. Indeed, because reviewers are asked to comment specifically on how well an applicant addresses these criteria, it can be helpful to the reviewer, and thus the applicant, to include within the proposal clear, concise, highlighted statements that address significance, innovation, and approach.

Because each reviewer can interpret what is meant by significant, innovative, or appropriate, it should not be surprising that different reviewers often have very different opinions with regard to the value of the same proposed research. Hence, the applicant should try to write in a way that engages different people having different backgrounds, noting in particular that not all reviewers are experts in the specific area proposed. Rather, it is expected that good scientists and engineers can recognize good work when they see it; a key challenge, therefore, is to help a reviewer to appreciate the overall objective and significance as well as the specific innovation and logic of the approach from both a general and a problem-specific perspective. This need may be even more important in cases where the reviewer has expertise in the proposed area but a different scientific opinion.

There are, nonetheless, many things that all reviewers appreciate. For example, because reviewers are often asked to review multiple proposals (e.g., 8–12), it is important to make the proposal easy to read and understand. Rather than using a small font or small figures to fit more information into the same space, use legible fonts (the NIH requires at least 11 point font, typically Arial, Georgia, Helvetica, or Palatino Linotype) and ample figures or tables to aid the reader; similarly, write clearly and concisely to increase the value of what is provided. Indeed, this brings us back to the importance of Chap. 2. Note, for example, that in a timeless 1987 publication of the NIH, entitled *Helpful Hints on Preparing a Research Grant Application to the National Institutes of Health*, one reads

> Try to develop a clear, concise, coherent scientific writing style. A few guidelines that may prove useful include: (1) using the active voice, which is more direct, less wordy, and less confusing than the passive voice; (2) keeping together related ideas and information, for example putting clauses and phrases in sentences as close as possible to the words that they modify; (3) simplifying and shortening overly long and involved sentences and paragraphs; and (4) eliminating redundant and awkward words, phrases, and sentences.

Additional subtle things can likewise make it easier for the reviewer. For example, although citing the many references numerically (often 100 or more) can save space and thus enable the applicant to include additional information, this forces an educated reviewer to refer back continually to References to determine what work has been cited. It can be constructive, therefore, to cite references by name and date, for example, Smith et al. (2021) or (Smith et al., 2021) rather than [20], for the reviewer may know of and respect the work of Smith and colleagues. By including dates as part of the in-line citations, one can also show the reviewers that the application builds on or advances current findings as well as foundational work from times past. Similarly, recall that schematic drawings, images, figures, and tables can literally be worth a thousand words if done well. Indeed, many reviewers will first skim through an application by paying particular attention to these visual aids. For this reason, one should provide descriptive legends so that the reviewer does not have to search the text to find meaning and importance. Because of the importance of the accompanying text, however, it is also good practice to refer to the schematic drawings, images, figures, and tables using boldface type to enable the reviewer to locate easily that part of the text that discusses each figure (e.g., boldface **Fig. 1** and **Table 1** are easier to identify quickly in the text than are Fig. 1 and Table 1). Although they should be kept to a minimum, introducing key well-accepted abbreviations in boldface also enables a reviewer to find them much more quickly in the text, as, for example, nitric oxide (**NO**) versus nitric oxide (NO). The key, therefore, is to keep the reviewer's perspective in mind at all times and not to compromise the use of effective devices and strategies for writing well because of stated limitations on the number of pages allowed—concise writing will generally provide the extra space needed to include all of the necessary information.

> **Exercise 4.3**
> The NIH website, www.nih.gov, has specific instructions for writing a K-series grant application. Write a three-page summary that would be sufficient as an overview of the motivation, scope, and requirements for submission of such a grant application to the NIH.

4.4 The NIH R01 Grant

As noted previously, the R01 grant is but one of many funding mechanisms administered by the NIH. One can obtain information about the other types of grants from the NIH website (www.nih.gov), but we consider here the format and requirements for the R01 because it represents well how to design an effective application.

The R01, or single investigator grant, includes a cover page with administrative information, brief description of the research (Project Summary, 30 lines maximum), lay description (Project Narrative, 3 sentences maximum), list of primary personnel, Table of Contents, Budget and Budget Justification, a Biosketch (5 pages maximum) for each of the primary personnel, information on Resources and Equipment (i.e., the research infrastructure) that are available to the investigator(s), the main body of the application (Specific Aims, Research Strategy, and References), information on Human Subjects and Vertebrate Animals, if any, Data Management and Sharing Plan, and further administrative information. With the exception of the main body of the application, much of this information must be provided within appropriate NIH-supplied form pages. Again, see the NIH website for instructions and details on the overall grant application package, including form pages.

Here, we focus on the main body of the R01 application, that is, the sections that detail the scientific need and elaborate and justify the proposed approach. At the time of publication of the 1st Edition of this book (in 2009), these sections included Specific Aims (1 page), Background and Significance (3 pages recommended), Preliminary Results (6 pages recommended, but no more than 9 pages), Research Plan (15 pages recommended, but no fewer than 12 pages), and References (no page limits, but admonishment to include only relevant citations). That is, not counting the References section, the NIH allowed 25 pages of single-spaced text to detail and defend the primary technical content of the application. At that time the exploratory/developmental research grant mechanism, referred to as the R21, allowed 15 pages of single-spaced text for the primary technical content. In 2010, however, the NIH dramatically reduced the page limits for both the R01 and the R21 mechanisms, the former from 25 to 13 pages (1 page Specific Aim plus 12 pages from Significance through Research Plan) and the latter from 15 to 7 pages (1 page Specific Aims plus 6 pages from Significance through Research Plan). Many investigators who had submitted successful applications using the prior page limits initially felt that it would be impossible to convey the same information as effectively in essentially one-half the space. Yet, consistent with our emphasis in Chap. 2 on the advantages of writing for conciseness, most investigators soon learned that the reduced page limits actually forced them to write better proposals, ones that were clearer and more compelling because of the added conciseness.

In addition to changing the allowable number of pages, the NIH (also in 2010) consolidated a previously required section (Background and Significance) into a single section (Significance), eliminated a previously required section (Preliminary Results), and added a new required section (Innovation). Such changes both remind us to refer to current instructions before starting any new application and reveal insight into what the NIH currently values most. Indeed, in 2016 the NIH further required applicants to detail the "scientific premise" in Significance, to demonstrate "rigor and reproducibility," to consider "age and sex as biological variables" in Research Plan, and to add a new section on "Authentication of Biological and Chemical Resources." Given some confusion in the community over the precise expectations, the term "scientific premise" was changed in

2019 to "weakness in prior research." Given the importance of these new requirements, which seek to increase transparency, note that according to the NIH:

> *Scientific Rigor*—The strict application of the scientific method to ensure robust and unbiased experimental design, methodology, analysis, interpretation, and reporting of results (Demonstrate in Research Plan/Approach).
>
> *Rigor of Prior Research*—One should describe and discuss in both Significance and Research Plan (Approach) the strengths and weaknesses of prior research to show the need for the proposed research (to address prior weaknesses) and to establish foundations (build on prior strengths) upon which it will be based.
>
> *Biological Variables*—To ensure better inclusivity, the NIH had previously required that the Human Subjects section should properly include diverse populations. The newer requirements built on this to emphasize the need to include age and sex as biological variables within the Research Plan (Approach) in human, animal, and cell research.
>
> *Authentication*—Describe the methods used to ensure the identity and validity of key biological and/or chemical resources used in the proposed studies (new section). For example, one would need to document the particular antibodies (i.e., source) to be used in a proposed experiment and to verify that the antibody is specific for the intended use.

Most recently (in January 2023), the NIH further emphasized the need to ensure rigor and reproducibility by introducing another required form page that must be completed for each application. For convenience, this page, *Data Management and Sharing Plan*, is reproduced at the end of this Chapter as Appendix 1. This plan must now be submitted with any proposal that intends to gather scientific data. If these data include large-scale genomic data, an additional form page (Genomic Data Sharing Plan) must also be completed and included. Briefly, it is clear from the form that the NIH now requires investigators to specify the types and amount of data that will be collected and to present specific methods by which these data will be archived and shared. It is axiomatic that shared data sets allow independent verification, which is increasingly important as data sets grow in size and complexity. Moreover, given the rising costs associated with modern research, another motivation is to avoid unnecessary duplication, which promises to accelerate the research while managing costs responsibly. As discussed in Chap. 7, yet another motivation is to minimize fabrication and falsification of data—temptations to misrepresent data should decrease with increased scrutiny of the data. Finally, it is not just the data, but also the methods used to collect the data and the software that is often used to interpret the data that now need to be documented better and, where possible, shared. Again, such sharing promises to promote reproducibility and reduce redundancy, thus accelerating research.

Before discussing in further detail each of the three primary technical sections of the main body of the R01 grant, note that each section or sub-section should answer a specific question:

Specific Aims (1 page): What are you going to do?
Research Strategy (12 pages):
    Significance: Why is it important?
    Innovation: How is it novel?
    Research Plan or Approach: How are you going to accomplish the work?
References (no page limit): What are the key findings on which you will build?

As in any good technical document, the writing should flow logically from section to section and the applicant should reinforce the main ideas throughout. Similar to the situation of multiple authors writing a paper, when multiple investigators write different parts of a proposal, it is important for the principal investigator to ensure a consistent style throughout, including tense.

4.4.1 Specific Aims (1 Page)

The Specific Aims section is particularly important; it is one of the few sections that all members of a Study Section must read. The Specific Aims should thus capture each reader's interest, show the need for the proposed research, and detail the specific results that will be sought—all in one page. Different applicants use different formats, but a general approach to constructing an effective Specific Aims page is to begin with two or three short paragraphs that identify the overall need, objective, or long-term goal of the applicant(s) as well as the specific hypothesis or problems that will be addressed and why, then list the individual specific aims, and conclude with a brief paragraph that highlights the overall significance and innovativeness of the work. An example of a good closing sentence to Specific Aims is: "Successful completion of this project will ...."

There is no limitation on the number of specific aims that one can propose, but most R01 applications focus on three to four aims, each of which can include sub-aims. Just as it can be efficient to begin writing a technical paper by identifying the primary findings, so too it can be efficient to begin writing a proposal by first identifying the specific aims. Indeed, it is often useful to draft the Specific Aims page and have multiple colleagues provide feedback on the overall motivation and plan before beginning to write the remainder of the proposal.

Specific aims should be just that, specific. Moreover, construct these aims in a forceful way—to quantify, to determine, to design, to test, to develop, and so forth. Many applicants construct each specific aim to test a specific hypothesis; if this is the case, then ensure that the aim is testable. Although one can provide some indication in this first section as to how the aims will be accomplished (e.g., using a particular animal model or data from a particular clinical trial), it is best to focus on methods and approaches in the section on Research Plan.

4.4.2 Significance and Innovation (1–2 Pages Recommended)

In some ways, the Significance section can be the hardest section to write well. Whereas one may think of this section simply as a brief literature review and statement of the obvious (e.g., that the proposal is significant because a particular number of Americans experience the targeted disease or condition), it actually must be much more. Within the context of answering the question, "Why is this research important?," the applicant should critically assess the literature to show convincingly what is unknown and why this lack of understanding (including prior lack of rigor) continues to impede scientific advances, improvements in health care delivery, the development of better medical devices, and so forth. For example, we may know that a genetic mutation is responsible for a particular disease, but prior research has yet to elucidate how this mutation affects the activity of a particular type of cell or the structure of a particular tissue or organ. Similarly, we may know that hemodynamic factors give rise to a particular vascular pathology, but prior research has yet to uncover how the associated forces induce the changes in gene expression that ultimately cause the disease. Although identifying gaps in understanding will often require one to point out shortcomings in prior investigations of others ("weakness in prior research"), we should do this diplomatically.

When identifying key gaps in the literature (including lack of rigor in prior studies), one should show convincingly the need for the proposed specific aims. In other words, Significance should motivate and "set the stage for" the Research Plan. Yet, remember that not all reviewers will be intimately familiar with the specific area of research, hence also use this section to help the reviewers to appreciate better the importance of the identified gaps and the proposed methods for research. Indeed, in this way, Significance should lead naturally to Innovation. In this complementary section, one should emphasize the novelty of the different approaches and methods that will be used to address the gaps and thus to complete the Specific Aims. A challenge here, of course, is that some reviewers tend to think of novelty as a truly new, inventive approach or method whereas it may be that the novelty of your study comes primarily from applying established approaches or methods to the problem at hand, which is novel and promises to yield new insight. For example, you may not propose to develop a new method for single cell RNA sequencing, but you may propose to use such sequencing for the first time to understand a particular disease or you may propose to use information from such sequencing to inform a novel computational model. Regardless, Innovation must be established clearly and shown to overcome limitations or weaknesses of prior studies that have prevented the field from moving forward.

To aid the reviewers, the applicant should highlight key points in all sections, for example, by **boldfacing,** *italicizing*, or <u>underlining</u> key parts of the text, though not in such excess that it becomes distracting. Given the overall importance of both Significance and Innovation, but limited space in these sub-sections, successful applicants are generally very good at highlighting the potential importance and planned novelty throughout the proposal—for example, the last paragraph of Specific Aims, the sub-sections devoted to Significance and Innovation, and Rationale sections within Research Plan all provide appropriate opportunities to remind the reviewer of the significance of the work and the innovativeness that will enable success. The challenge, therefore, is to reinforce key points throughout without redundancy.

4.4.3   Research Plan (11–10 Pages Recommended)

Recall that the primary question that needs to be answered in this section is, "How are you going to accomplish the proposed work?" In conjunction with Specific Aims, this section is perhaps the most important and thus demands careful attention. The best words to remember when writing this section are "detail" and "clarity."

There is no required format for Research Plan, yet an effective strategy has evolved over the years. Many applicants begin this section with a paragraph that highlights the overall research plan and its importance, sometimes including a schematic drawing to show how the different aims complement one another. Next, they describe the Rationale, Methods, and Expected Results and Potential Limitations for each aim in sequence. Finally, they conclude with a brief summary of the overall project and an expected timeline to accomplish the project. One small variation on this strategy has also arisen in recent years, due in large part to the complex but common procedures used in molecular and cell biology. Similar to the format of some technical journals, one can collect detailed methods (often common to multiple aims) at the end of this section, almost like supplemental materials, so as not to interrupt the flow of the main portion of the section; this strategy allows the interested reader to evaluate the appropriateness of the details after having focused on the scientific questions to be addressed.

If one adopts the most common strategy, then the basic outline for the main portion of Research Plan becomes either[2]:

Aim 1. Restate the specific aim exactly from the first page.
Rationale.
Methods.
Expected Results & Limitations.

---

[2] There are many slight variations, however. For example, one could replace the generic methods section with separate sections on experimental design and data analysis..

Aim 2. Restate the specific aim exactly from the first page.
Rationale.
Methods.
Expected Results & Limitations.
Aim 3. Restate the specific aim exactly from the first page.
Rationale.
Methods.
Expected Results & Limitations.

or,

Aim 1. Restate the specific aim exactly from the first page.
Rationale.
Methods.
Expected Results.
Potential Difficulties & Alternate Strategies.
Aim 2. Restate the specific aim exactly from the first page.
Rationale.
Methods.
Expected Results.
Potential Difficulties & Alternate Strategies.
Aim 3. Restate the specific aim exactly from the first page.
Rationale.
Methods.
Expected Results.
Potential Difficulties & Alternate Strategies.

Restating each aim in its entirety reminds the reviewer of the specific goals—to quantify, to determine, to design, to test, and so forth. Restating each aim in boldface serves as a natural and effective visual cue for organizing this long section; this approach is much less distracting than needless section numbers such as 3.1, 3.1.1, and so forth. Remember that these Aims should be complementary, but not co-dependent. In other words, reviewers often ask themselves "what if one Aim fails," does this mean that all subsequent Aims will similarly fail. There is, therefore, a need for a careful balance in constructing and communicating these Aims. Whereas the Significance section discussed previously should focus primarily on the importance of accomplishing the overall project and the Innovation section should highlight the novelty of the methods and overall study, Rationale should focus on the fundamental motivation(s) for each specific aim. For example, the applicant should note what important gap in our understanding this specific aim will address and why the adopted approach is innovative.

The methods section for each aim is similar to a methods section in an archival paper—it should provide methodological details sufficient to enable the reviewer to understand and appreciate fine details of the study. For example, one should not write "the cells will be cultured in an appropriate media." Rather, one should document the specific media to be used (including vendor), any supplements with appropriate concentrations, the temperature and CO_2 level, and how often the media will be refreshed. Similarly, one should not write "the governing differential equation will be solved numerically" or "the data will be analyzed for possible statistical significance." Rather, one should provide details on the specific numerical method and why it is appropriate for the expected class of differential equations, and similarly one should provide details on the specific statistical tests, including post hoc testing, why they are appropriate, and the levels of desired significance. Again, the operative word in this section is detail, assuming of course that the methods are appropriate and proven. If the methods are new, one should provide even greater detail. With the availability of web-based archives such as Cold Springer Harbor's bioRχiv, it can be prudent to deposit a detailed citable (doi) paper when the methods are novel and complex and not yet supported by peer-reviewed publications.

Although NIH-funded research does not need to be hypothesis-driven, one should always anticipate the results. It is thus prudent to discuss why you expect such results, which actually allows one to justify further the importance and implication of the aim. Although results should be new, it is always good to cite related studies that provide further confidence that the aim will prove successful and important (this helps to address the call for rigor and reproducibility). Similarly, although methods should be chosen and justified carefully, it is always possible in science and engineering for difficulties to arise that prevent one from conducting the experiments or analyses as originally planned. There is also a need, therefore, to anticipate potential difficulties and to have reasonable contingency plans. Just as in the discussion of an archival journal paper, however, one must achieve the proper balance in identifying potential pitfalls while not implying that the aim will be very difficult to achieve as planned. It is wise to discuss the presentation of this balance with a valued colleague.

Finally, it is useful to conclude the Research Plan with a detailed timeline showing the anticipated duration of each part of the project and how the different parts will progress together, sometimes in different laboratories. It is also good to provide a single paragraph that concludes the grant—remind the reviewer that the innovative approaches promise to yield significant findings that will advance the field.

4.4.4   References

The reference section for a R01 grant was once limited to four pages, but there is currently no such page limitation. Nevertheless, one should not seek to compile an exhaustive list of references; it is more important to be selective, focusing on the key papers that support the need for the research and the methods that are appropriate to address this need. Similarly, there is no required format for references except that each must include the list of authors, year of publication, title of the work, the publisher (if a book), or journal title, volume, and inclusive pages. Because reviewers are typically familiar with the proposed research area, and thus the key authors in the field, it can be helpful to list the references alphabetically. Indeed, this is consistent with the aforementioned recommendation to cite by author (e.g., Smith et al., 2021) rather than by number (e.g., [20]), for this eliminates the frustration felt by knowledgeable reviewers who do not want to go back and forth to the references to see who did what. Because of the availability of research papers through the Web, some applicants also provide the doi to enable the reviewers to download key references easily.

4.4.5   The R01 Summarized—Stay Current

We provide in Appendix 2 a brief summary in the form of a checklist to help the reader to prepare an effective NIH application and provide in Sect. 4.5 below additional strategies to consider when preparing an application. In concluding this section, however, we cannot overemphasize the need to visit the website(s) that describe a particular funding opportunity for the associated goals and requirements change frequently.

For example, ongoing discussions at the NIH may lead to changes in the criteria used to score applications. Many of these discussions are currently motivated by the desire to remove potential bias from reviews. As noted above, the five primary criteria that are currently scored (from 1, exceptional, to 9, poor) are: Significance, Innovation, Approach, Investigators, and Environment. It has been suggested that these five criteria be assessed with the framework of three factors:

Factor 1—Importance of the Research (Significance, Innovation), scored from 1 to 9.
Factor 2—Rigor and Feasibility (Approach), scored from 1 to 9.
Factor 3—Expertise and Resources (Investigator, Environment), assessed and considered within the Overall Impact Score (1–9), but not scored individually

---

Exercise 4.4
Investigate past changes in addition to this possible change in the way NIH grants are assessed and scored and write a three-page summary, including potential advantages and disadvantages of such changes.

4.5   Additional Strategies for Success

A successful proposal excites reviewers with its potential impact and convinces them that the project is likely to succeed or, at least, that the potential benefits of success are worth the risk. Thus, mastering the mechanics of writing a proposal is only the beginning.

Presenting your work in the right way to the right audience (review panel) is also essential. This section highlights some important tips that the authors have learned during their own careers from writing and reviewing grants.

4.5.1 Feedback, Feedback, Feedback

Many scientists and engineers still remember the first time that they submitted a grant. Months of tiring work but a sense of accomplishment once the submission was complete—and then nothing for a long time. Depending on the agency and specific grant program, the time between submitting a proposal and receiving written reviews is often six months or more. This delay has a very important implication for new faculty members; refining a proposal after waiting for reviews, resubmitting, and repeating this process multiple times is a slow way to build a laboratory and establish a funding portfolio.

By contrast, asking a senior colleague having experience reviewing similar proposals to read and comment on a draft of a grant requires just a couple of weeks. In our experience, a mock grant review by experienced colleagues is the most effective way to improve a proposal, increase its likelihood of getting funded, and shorten the total time required to secure funding for a project. Recognizing the power of this simple idea, many universities organize mock reviews for their faculty through the offices of the Vice Provost for Research or Associate Dean for Research. Amazingly, these services are typically underutilized because many faculty are unwilling to produce a full draft of a grant proposal 4–6 weeks prior to the submission deadline—in other words, they are unwilling to finish a month early to save a year! If your institution does not provide this service, former mentors are often willing to help; there are also ideas in the next section for identifying potential mock reviewers both within and outside your institution.

Feedback from experienced colleagues at one point in the writing process is good; feedback throughout the process is even better. While it may seem intimidating to share ideas that are not fully developed, early feedback is essential to refine ideas and ensure that time invested in assembling preliminary data and proposals is well spent. When writing an NIH proposal, we recommend seeking feedback on the Specific Aims page as soon as you can produce a solid draft. The Specific Aims page concisely explains the significance of the problem, the work that will be done, and the impact if it is successful. The Aims page sets the tone for the entire proposal; if reviewers are not excited after reading this page, the review will not go well. Yet it takes only a few minutes for an experienced colleague to critique this single page, making it possible to get feedback from multiple people and to iterate multiple times before embarking on writing the full proposal. Another similar tactic is Grant Brewing sessions, where scientists pitch their Aims or Objectives to a room of colleagues for real-time feedback.

There is one more point in the grant-writing cycle where feedback from experienced colleagues is essential—interpreting and responding to reviews. It takes time to learn how to read reviewer comments, to understand which are most important to address, and to decide how to address them. Each agency has its own review formats, jargon, and scoring systems, which a colleague who reviews for that agency can help navigate. For example, at the NIH, strong scores for Significance and Investigator suggest that a revised proposal with an improved approach might fare well. By contrast, a grant with low Significance or Innovation scores might face a more difficult path, regardless of how well other aspects are rated on the current or future submissions.

Resubmitted NIH proposals must include an additional one-page section, an Introduction. It can be just as important to receive early feedback on a draft of this page as it is for the Specific Aims page. In this Introduction, the applicant should (if applicable) restate strengths noted previously by the reviewers (e.g., that the work is significant, that the investigative team is excellent, and so forth) but mainly list the noted weaknesses and how they have been addressed in the resubmission. In cases where a reviewer clearly misunderstood an aspect of the proposal, assume that you were not clear and work to increase clarity. In cases where the reviewer was not convinced that the proposed experiment or computation would be successful, generate additional pilot results (and archive on a citable preprint server if possible) that demonstrate feasibility and utility. In cases where a reviewer simply disagrees, try to identify studies by other investigators that better support your hypothesis or approach (which increases rigor). Given that NIH only allows one resubmission, it is better to be highly responsive than to be quick to resubmit, even if takes months to generate appropriate pilot results. Another key to success in this regard is to be willing to rewrite much of the proposal; trying to retain much of the prior text is often not effective.

4.5.2 Know (and Seek) Your Audience

In most situations, the agencies and foundations who fund scientific research can afford to fund only a fraction of the applications they receive. To be selected, a proposal must not only be of high quality, it must also capture the attention of the reviewers or officials making the funding decisions. Thus, an underappreciated aspect of grant-writing is getting your grant in front of an audience that is likely to be excited by it. We have seen many grants rejected by one review panel only to be scored very well by another, often within the same granting agency.

The first step in understanding your audience is to gather information about how decision-making works at agencies and foundations that are potential funders. As we noted above, review is a two-step process (Study Section and Council) at the NIH, with Program Officers not involved in the first step. By contrast, individual Program Officers for some Department of Defense programs have considerable latitude to choose the projects that best fit their mission. In those cases, getting to know the appropriate Program Officer(s) and developing concepts through interactive discussions is likely the most important step an investigator can take to improve the odds of success. At the National Science Foundation, the situation falls somewhere between these two examples. Panels of reviewers rank proposals, but the Program Officer has some discretion in selecting proposals based on those rankings. In this situation, discussing ideas with Program Officers in advance to understand the goals of a particular directorate and program is an important part of finding the right audience.

Returning to the NIH as our illustrative example, the NIH RePORTER database of funded grants is a valuable resource. This database allows scientists and engineers to identify grants that have been funded in an area of interest, who is leading those projects (a potential source of mock reviewers within or outside your institution), and which Study Sections reviewed each of those grants. Exercise 4.5 guides the reader through a series of steps to identify potential mock reviewers and Study Sections, starting with at least a title and preferably a draft Specific Aims page for an NIH R01 or R21. The exercise includes steps to identify one or more possible Study Sections when you submit your application. Actually getting a grant assigned to a particular panel depends, however, on the mechanism by which those assignments are made, which evolves over time.

Exercise 4.5 also includes steps to identify colleagues working in similar areas who might be willing to provide a mock review. Asking colleagues outside of your institution to review a grant draft can carry some risk, however, for several reasons. First, never approach a member of a Study Section directly for advice; this is considered a serious breach of the review process and must be reported to the NIH. Second, most Study Sections recruit additional experts on an ad hoc basis who are not listed on the website, depending on which applications have been submitted. If colleagues who like your work have already helped you with a grant and then are invited to join a panel, they would have to recuse themselves, costing you a potentially receptive reviewer. Finally, you should only share preliminary proposals with those whom you trust to keep the information confidential. For these reasons, when possible, it is safer to work with colleagues at your own institution or those with whom you have a close professional relationship (former mentors, current collaborators) that would already preclude them from reviewing your grant.

Exercise 4.5
Starting with at least a draft title and preferably a draft Specific Aims page for an NIH grant you are beginning to write, use the following steps to target a Study Section and generate ideas for potential mock reviewers.

1. Go to the NIH RePORTER website and create an account so that you can save searches and return to them.
2. Use the Text Search field to search for some keywords related to your project in the Titles, Terms, and Abstracts of active (funded) projects. (You can also use the

Fiscal Year field to specify dates if you want to find previously funded projects that are not currently active.)

3. Scan the list. Do these titles sound similar to your project? If not, refine your keywords and search again. Write down the names and institutions of the PIs you recognize; any you know well—especially at your own institution—might be good candidates to provide a mock review of the proposal, with the caveats listed above.

4. Next, pick a project that sounds interesting and potentially related and click on the title; this will bring up a screen with the project abstract and other details about the project, including the Study Section that reviewed the proposal.

5. Browse your list and identify the Study Section(s) that associate with the largest number of projects. It may be easier to use the Export feature of RePORTER to download selected information for the proposals into a spreadsheet that you can use to sort and search.

6. Now, go to the NIH Center for Scientific Review (CSR) website and find the Study Section(s) you identified. Do the topics listed look like a good match for your project? If not, try other Study Sections from your list of refine your original keyword search.

Finally try two other cross-checks to decide if you have found a good Study Section. First, use RePORTER to generate a list of all active projects reviewed by that Study Section, and see if they look like a good fit. Second, review the lists of panel members from the past few meetings; ideally you should recognize names of some people who work in your field, attend the same conferences, and publish in the same journals.

4.6   The Preproposal

Perhaps the best example of the need to write concisely with clarity is the preproposal. Because of the greater numbers of applicants applying for limited financial resources, many agencies have instituted a two-stage review process. The applicant(s) must first submit a brief preproposal to be reviewed by an expert panel. Based on the findings by this panel, only a subset of full proposals is invited for further review and consideration for funding. A good example of a two-stage review process in biomedical research is that used by the Leducq Foundation (www.fondationleducq.org). For example, this foundation may review on the order of 100 preproposals per year, then based on initial reviews invite only on the order of 10 applicants to submit a full proposal from which they may select three to four proposals for funding.

The state of Texas, as another example, had a competition called the Advanced Research Program (ARP) that was open to any full-time member of the faculty of a

Texas institution of higher learning. Prepreposals for ARP grants were limited to 4000 characters (use the word/character counting feature of your word processor to count); this is essentially 1.3 pages, double-spaced, in 12 point font—not a lot of information. Yet, a panel would decide whether to invite the applicant to submit a full proposal, the next important step toward possible funding, based solely on these 4000 characters. Again, the need for clarity and conciseness is clear. The format for the ARP prepreposal was simple:

Project goals and methods
Staff
Facilities and resources
Education and training.

Although such prepreposals are short, one clearly wants to communicate information similar to that contained in the much longer R01 application: What are you going to do? Why is it important? Are you capable of being successful? How are you going to accomplish the work? Thus, the key thing when beginning a proposal of any type is to determine the information that will best represent you and your ideas. Only then will you be able to decide best how to package this information within the format for the particular agency.

4.7   Closing Comments

Another foundation that funded biomedical research was the Whitaker Foundation, which closed in 2006. Over many years, however, it provided millions of dollars of funding to support new investigators in biomedical engineering and to develop new academic programs. They provided reviewers of individual investigator grants with a checklist to ensure that applicants covered a number of critical aspects of research in biomedical engineering. Reasons for scoring a Whitaker application poorly included:

No clear hypothesis
Mundane/uninteresting
Little engineering
Little biology
Not enough detail
Unrealistic/faulty approach
Needed collaboration missing.

Other reasons commonly cited for scoring NIH applications poorly include:

Not significant; not innovative; not exciting
Unjustified hypotheses
Unaware of previous related work
Insufficient pilot data
Poorly designed research plan; unorganized
Overly ambitious
One or more aims are poor
The success of one or more aims depends on the success of a previous aim.

Although we should focus on the positives, it is prudent to appreciate causes for failure. In summary, some of the most important reminders for grant writing are:

Know the mission of the agency and target the proposal accordingly. For example, you would not think of sending a proposal on cancer research to the American Heart Association.
Read the instructions and follow them carefully when preparing your application.
Read a recently funded proposal to the agency to which you are applying.
Ensure that the proposal addresses an important issue and offers the potential for significant advancement; get early feedback on the Specific Aims.
Remember that your proposal must generally address simultaneously two technical audiences: those who are very familiar with the field and those who are less so.
Finally, finish early so that colleagues can review the application and provide constructive criticisms that you have time to employ. Only in this way can we avoid the common pitfalls that plague so many proposals.

Exercise 4.6
Write a 4000-character preproposal using the Texas ARP format. Select a topic of interest to you, assume that you are the only investigator, and describe resources available in your laboratory or department that would be sufficient to conduct the work.

Exercise 4.7
The NIH website (nlm.nih.gov/ep/Tutorial.html) provides useful guidelines on How to Write a NIH Grant. Go to the site(s), review the material, and prepare a 25-slide PowerPoint presentation that could be used as an introduction to writing and submitting NIH grants.

Appendix 1: New NIH Data Management and Sharing Plan

The interested reader is always referred to the NIH website because requirements change frequently. Nevertheless, below we reproduce from the current (February 2023) web-site the six required elements for this new required form page.

Element 1: Data Type

A. **Types and amount of scientific data expected to be generated in the project:**
   Summarize the types and estimated amount of scientific data expected to be generated in the project,

B. **Scientific data that will be preserved and shared, and the rationale for doing so:**

   Describe which scientific data from the project will be preserved and shared and provide the rationale for this decision.

C. **Metadata, other relevant data, and associated documentation:**

   Briefly list the metadata, other relevant data, and any associated documentation (e.g., study protocols and data collection instruments) that will be made accessible to facilitate interpretation of the scientific data.

Element 2: Related Tools, Software and/or Code
State whether specialized tools, software, and/or code are needed to access or manipulate shared scientific data, and if so, provide the name(s) of the needed tool(s) and software and specify how they can be accessed.

Element 3: Standards
State what common data standards will be applied to the scientific data and associated metadata to enable interoperability of datasets and resources, and provide the name(s) of the data standards that will be applied and describe how these data standards will be applied to the scientific data generated by the research proposed in this project. If applicable, indicate that no consensus standards exist.

Element 4: Data Preservation, Access, and Associated Timelines

A. **Repository where scientific data and metadata will be archived:**

   Provide the name of the repository(ies) where scientific data and metadata arising from the project will be archived).

B. **How scientific data will be findable and identifiable:**

Describe how the scientific data will be findable and identifiable, i.e., via a persistent unique identifier or other standard indexing tools.

C. **When and how long the scientific data will be made available**

Describe when the scientific data will be made available to other users (i.e., no later than time of an associated publication or end of the performance period, whichever comes first) and for how long data will be available.

Element 5: Access, Distribution, or Reuse Considerations

A. **Factors affecting subsequent access, distribution, or reuse of scientific data:**

NIH expects that in drafting Plans, researchers maximize the appropriate sharing of scientific data. Describe and justify any applicable factors or data use limitations affecting subsequent access, distribution, or reuse of scientific data related to informed consent, privacy and confidentiality protections, and any other considerations that may limit the extent of data sharing.

B. **Whether access to scientific data will be controlled:**

State whether access to the scientific data will be controlled (i.e., made available by a data repository only after approval).

C. **Protections for privacy, rights, and confidentiality of human research participants:**

If generating scientific data derived from humans, describe how the privacy, rights, and confidentiality of human research participants will be protected (e.g., through de-identification, Certificates of Confidentiality, and other protective measures).

Element 6: Oversight of Data Management and Sharing:
Describe how compliance with this Plan will be monitored and managed, frequency of oversight, and by whom at your institution (e.g., titles, roles).

Appendix 2: (Copy and Use This as a Quick Reference)

Specific Aims (What you are going to do?)

- The first sentence or two should engage the reader and motivate the need for the work.
- Briefly note long-term goals/overall hypotheses, then focus the work.
- State your Specific Aims (three to four) and how you will achieve/test them.
- Conclude by emphasizing the significance and innovation of the proposed work.

Significance (Why it is important?)

- Significance refers to overall importance and long-term potential rather than the significance of individual aims—address the latter as rationale in Research Plan
- Review the literature critically, that is, identify foundations as well as weaknesses in rigor of prior work. Do not simply state that A did this, B did that, and C did that. Identify weakness in prior research that your work will address and resolve.
- Being unaware of important findings in the field does not engender confidence; conversely, citing current literature and presentations from recent meetings by leaders in the field suggests that you are on the cutting edge (do not overdo though).
- Only a few of the reviewers will have expertise in the specific area, yet many will read the proposal. This section should educate the general reader.

Innovation (How is your work unique?)

- This section should accomplish two things: demonstrate your capability in the research area and highlight the unique approaches, methods, questions, expected results that promise to advance the field.
- Seek a balance that engenders confidence that the work is feasible, exciting, and important, not overly ambitious or risky (novel methods can be viewed as untested).

Research Plan (How you are going to accomplish the work?)

- One of the most effective strategies is to address each aim separately, but to do so in a consistent, well-ordered manner. For each aim, use subsections, as, for example, (a) Rationale, (b) Methods, (c) Expected Results & Limitations.
- The Rationale of each aim should address the importance of this part of the project and how it fits into the overall/long-term goal. This is also a good time to remind the reader of novelty or innovation. One short paragraph should suffice.
- Methods for each aim may include materials, equipment, theoretical frameworks, assays, statistical methods, and so forth, all given in sufficient detail. For example, do not merely say that a physiological solution will be used—give the specific composition. Similarly, do not just say that a particular device will be used—give the resolution of the device and any unique capabilities.
- Whether hypothesis- or curiosity-driven, one should know what to expect with regard to findings. Discuss this and note its potential importance. Likewise, one should recognize limitations and anticipate pitfalls that may arise while offering possible alternate strategies. Noting and addressing possible limitations is much better than hoping that a reviewer will not think of them; someone always does and this could relegate an otherwise outstanding proposal to a lower score.
- Remember that detail is the operative word in Research Plan and that Specific Aims should form a logical, supporting sequence but not depend one on the other. You do not want a reviewer to discover that if one aim fails that the entire project is in jeopardy.

Oral Communication and Posters

5

Just as we must write well, so too we must speak well—a belief that is not new to modern science or engineering. According to Boorstin (1983, p. 395), Bishop Sprat suggested that the goal of the Royal Society of London (founded ~ 1660) was "not the Artifice of Words, but a bare knowledge of things." Hence, they

> extracted from all their members, a close, naked, natural way of speaking; positive expressions; clear senses; a native easiness: bringing all things as near the Mathematical plainness, as they can: and preferring the language of Artizans, Countryman, and Merchants, before that, of Wits, or Scholars.

In other words, as Boorstin concluded, "It was not enough that the language of science be simple. It had to be precise—and, if possible, international." Although audiovisual aids available today are very different from those of the 17th century, the need for simple, clear, and informative presentations remains.

Written documents and oral presentations both reflect one's professional reputation. Yet, the oral presentation is unique in that it can serve as the all-important "first impression." If a talk is lucid and enjoyable, those in the audience will likely seek out the speaker again; if a talk is poorly organized and boring, it may be the last time that they seek to hear the speaker.

Exercise 5.1
The need for excellence in oral presentations is not unique to science and engineering. Hence, find a good book on public speaking and read two chapters that are particularly appealing. Write and submit a three-page summary of the main points.

© The Author(s), under exclusive license to Springer Nature Switzerland AG 2024   91
J. D. Humphrey and J. W. Holmes, *Style and Ethics of Communication in Science and Engineering*, Synthesis Lectures on Engineering, Science, and Technology,
https://doi.org/10.1007/978-3-031-39125-5_5

Among the many books available, consider the timeless work, *How to Develop Self-Confidence and Influence People by Public Speaking*, by Dale Carnegie.

5.1   Effective Styles

Carnegie (1956) suggests that four things are essential in one's pursuit of becoming an effective public speaker:

1. Start with a strong and persistent drive.
2. Know thoroughly what you are going to talk about.
3. Act confident.
4. Practice, practice, practice.

Although these four essentials should not surprise anyone, they should cause some reflection. In particular, just as in writing well and ensuring integrity in the workplace, effective oral presentations do not just occur, even with experience—one must resolve to learn to present well and to continue to improve. Moreover, because of the importance of self-confidence when speaking to either small or large audiences, it is essential to know the subject so well that you could give the talk even if the audiovisual equipment failed or if you forgot your typewritten notes. Finally, the old adage "practice makes perfect" is certainly true, but there is one caveat. One can practice a bad talk over and over, but it need not improve. Rather, one's practice should include constructive criticism by peers that focuses on both the technical material and the method of presentation; it is better to make mistakes among friends and to receive helpful suggestions or corrections before the actual presentation.

Exercise 5.2
Attend three professional seminars and record seven specific personal habits and seven audiovisual techniques used by the speakers that were particularly effective (four each) or ineffective (three each). Summarize your findings in a table and submit a two-page report.

Many suggest that much of communication is nonverbal during discussions between individuals. Do we look the other person in the eye and reveal our interest or do we look at other people or things while they are talking? Do we change our facial expressions appropriately to reveal empathy or understanding or do we remain stoic? So too in public speaking, nonverbal communication can help make a talk engaging or it can render the attempt boring or, even worse, annoying. By definition, habits are natural and repetitive;

they usually arise unconsciously and can manifest nonverbally or verbally. For this reason, it is essential to have peers provide feedback on potentially distracting habits that arise while we speak. For example, if one tends to jingle keys in their pocket when nervous, recognizing this problem allows them to remove the keys before speaking, thus removing the potential distraction. Similarly, if one uses a lot of *aaahs* or *uumhs*, there is a need to identify these problems and to remove them from both formal and informal speech, for we develop new habits through consistency. Indeed, we have found that paying careful attention to composing well-written documents also serves to help us speak well. Finally, if one's hands shake badly during a talk, it is best not to use a laser pointer, which will project exaggerated motions onto the screen. Instead, one whose hands always shake should practice using verbal cues such as "as seen in the first term of Eq. 1" or "as illustrated well in the top curve in the left panel." A laser pointer can be an effective aid if used well, but it can also be very distracting. Indeed, even if held by a steady hand, a rapidly moving or constantly circling laser pointer can be a significant distraction. Finally, be careful not to keep the laser on if you "talk with your hands," for the audience gets both distracted and concerned when the laser shines across someone's face or constantly goes from floor to ceiling.

Valiela (2001) correctly suggests that effective technical presentations share some commonalities with successful theater. Two prerequisites for good theater are a good story and actors who "connect with" or "relate well to" the audience. A good story in science or engineering requires an interesting or important problem to be formulated, then solved in a novel and logical manner. Below, however, we focus on relating the story well to an audience, first by tabulating reminders related to basic techniques and habits of effective presentations. Indeed, although it is essential in science and engineering to have something important to say, as you compare the suggestions below, consider the suggestion of Carnegie (1956) that "It is not so much what you say as how you say it."

DO	DO NOT
Be confident, appear confident	Be arrogant or prideful
Be enthusiastic—it is contagious	Pace too much
Speak loudly, clearly, slowly	Speak in a monotone voice
Be respectful of questions	Ask rhetorical questions
Finish early enough for questions	Go over the allotted time
Maintain balanced eye contact	Look only at the screen or at a distance
Dress appropriately	Apologize for dress
Use (laser) pointer effectively	Circle everything with laser pointer
Know your audience	Discourage interactions
Define terms, use analogies	Use jargon, try to impress
Minimize nervous habits	Assume every talk begins with a joke

Exercise 5.3

Why is it important to be or at least appear confident? Why is it important not to be prideful or arrogant? What message will we convey to an audience if we finish early and allow questions? What message will we convey if we go over the allotted time and ignore calls to stop? What is the appropriate dress for different audiences? Ask yourself these and similar questions regarding this tabulated list of things to "do" and "not do," and write a two-page summary, including additional things to do or to avoid. If possible, conclude the summary with a few overarching statements.

Experienced actors tend to be nervous on opening night, and so too experienced speakers before walking up to the podium. Yet, recognizing that nervousness is natural, indeed expected, allows us to identify ways to minimize its effects and to settle quickly into a comfortable rhythm. For example, an early visit to the room where you will speak will help you to feel more at ease—the environment will not be foreign. If you need to use a microphone and have not done so before, ask the A/V technician if you can test the system before your talk. If you expect to be nervous nonetheless, eat sparingly before the talk to avoid further complications of the nervousness. Having complete command of the technical material will also engender self-confidence, which is the best way to negate nervousness. Remembering the first sentence or two will ensure a good start, which is essential in transitioning from nervousness to confidence. Beginning with an engaging slide will capture the audience's attention, which will reinforce your confidence (provided that you make eye contact with the now engaged audience). Conversely, memorizing a talk word for word can promote nervousness; you may become concerned that you will forget something and lose track of your message or even worse you will begin to stumble if you forget your next line. Below we tabulate some reminders related to basic techniques that promote confidence as well as contribute to telling the story well.

DO	DO NOT
Visit the room before speaking	Show up late or just before your talk
Remember the first sentence	Memorize the talk or read it directly
Use an engaging first slide	Start with a bulleted outline
Use slides as your reminders	Require audience to read a lot on their own
Maximize good figures/images	Use lots of words and small fonts
Be consistent in slide format	Mix slides with different backgrounds
Use slides to capture attention	Use slides to communicate most of the information
Remember your concluding remarks	End by saying, "Well that is all I have"

Reflecting on these suggestions, it should be clear that we recommend that one *use audiovisual aids to support a talk, not to carry it.* In other words, speakers should strive to capture the audience's attention so that they look at them and only look to the slides when so directed for clarification. Hence, the speaker should use comments like "and thus x is important, as illustrated well in this figure" or "x..., as can be seen in this image," noting that a well-used pointer can remind the audience when and where to look. Conversely, detailed text on a slide will usually entice the audience to read on their own and not look at or listen to the speaker; this situation should be avoided. Use slides primarily to show clear black and white or color images and figures, schematic drawings and flowcharts, equations, and to a lesser degree, tables, each of which should support what is said. Providing a short heading on each slide can indicate the focus of that slide; beginners may also put bullets on the slides as further reminders to themselves, particularly to prompt appropriate transitions. When referring to figures, start by defining the variables of interest and the axes; when referring to equations, start by defining the meaning of important variables or terms; when using color images, use the different colors as indicators of important features or points. Remember, however, that a portion of your audience will likely be color blind, hence avoid color combinations that they will not be able to distinguish. Remember, too, that less information explained well is always better than more information explained poorly.

Software programs such as PowerPoint can be tremendous tools when used well. Resist the temptation, however, to use all the "bells and whistles." For example, having figures fly in from the edges of a slide or animating molecules that come to screeching stops generally distract from technical content. Similarly, using complex backgrounds, particularly ones with gradients in color, can be less effective overall—some words show up well, while others do not. Depending on the fixed lighting in the room, slides having dark backgrounds can excessively darken a room and thereby create a more conducive environment for sleeping. For these and other reasons, black print on a white background and color images on a white background continue to be effective for they generally project well, maintain modest lighting in the room, do not discriminate unnecessarily against color blindness, and even allow one to use information directly from print versions of abstracts, proceedings, or papers that appear primarily in black and white because of considerations of cost. Regardless, consider adopting a common format/master slide, which enables you to interchange slides from different talks with minor modifications; having a common format (including font sizes for headings versus text) enables you to insert any slide from a different talk into the present talk without modification.

The first slide is traditionally a title slide—it should include a brief (60–120 characters) but informative title and list the authors and their affiliations. As appropriate, disclosures can be noted on the bottom of the first slide. Many try to add a touch of color to the opening slide by showing the university or business logo or perhaps a picture of a building or scenic area in which the group works. The last slide is traditionally an acknowledgment slide—it should list others who contributed to the work and financial support. Some prefer to read the names and the funding agencies, but it is often sufficient simply to list them and let the audience read for themselves. The last slide often remains projected the longest, that is, during the question and answer period, thus it is also a good place to list key references to your work and to provide contact information (e.g., an e-mail address). By contrast, the next to the last slide usually provides a summary of the work or the "take home" message. It is best to end on a high note, emphasizing the major findings, not to list all of the limitations or future needs. Address such needs in response to appropriate questions. If the presentation is for a job interview or request for funding, it can be effective to advance past the acknowledgment slide to a final conclusion slide to highlight points that you would like the audience to remember or to ask about.

Exercise 5.4
Prepare a 15-min PowerPoint presentation on effective grant writing. Practice the talk, paying particular attention to the time limit. Have two or three peers critique the presentation, then make corrections and repeat the presentation.

Exercise 5.5
Prepare a 15-min PowerPoint presentation on a technical topic of your choice, but do so in a way that that highlights bad presentation skills and personal habits. For example, use different backgrounds from slide to slide, use small fonts, use long detailed quotes, read directly from the slides, and so forth. Exaggeration often provides an important reminder of what not to do.

Exercise 5.6
Prepare a 15-min PowerPoint presentation on a topic of your choice that addresses an issue having potential ethical consequences. For example, previous students in our classes have discussed embryonic stem cell research, cloning, the use of human subjects in clinical trials, animal research, issues of science and religion, patents, and copyright.

5.2   The 15-min Presentation

Seeing your name appear in print on a journal article generally produces a sense of accomplishment and pride. So too, learning that your abstract or paper has been accepted for a podium presentation at a professional meeting produces a sense of excitement. After the initial euphoria, however, you realize that you have to find a way to describe in a short period, often 12–20 min, a project that you may have worked on for months or years. One is tempted, therefore, to pack as much information into the talk as possible. Surely the audience will be impressed by how much you did, right? As noted above, however, you will generally make a much more positive impression if you present less information well. There is, therefore, a critical need to identify the most important information and to ensure a logical sequence from identifying the problem to interpreting the results and appreciating the significance. Similar to writing a technical paper, a good way to start this process is to collect together all of the figures, images, equations, tables, and other major findings that you may include, then prioritize and order them in the most logical fashion. Again, this ordering need not be chronological; in many cases it is best to order the talk in the way that makes the most sense in hindsight.

A good rule of thumb is to prepare approximately one slide per allowed minute of presentation, including the first (title) and last (acknowledgment) slides, which need not be discussed. Moreover, each slide should generally highlight one main idea. Again, we emphasize that the first slide after the title slide should capture the audience's attention. It is much more effective, for example, to show a picture or image that motivates the work than to show a bulleted outline noting that you will introduce the overall problem, describe some of the methods, discuss the results, then draw conclusions—one expects such an approach. Carnegie (1956) suggests multiple ways to capture the audience's attention immediately: "arousing curiosity, relating a human interest story, beginning with a specific illustration, using an exhibit, asking a question, opening with a striking quotation, showing how the topic affects the vital interest of the audience, or starting with a shocking fact."

Consider here a brief anecdote that highlights the importance of the second slide (or first slide when one does not use a title slide) in a PowerPoint presentation. One of the authors was asked to give the second technical talk at an anniversary celebration for a College of Engineering. The first technical talk followed directly some brief comments by the President of the university. Out of courtesy, the President remained for the first talk because it began immediately following his comments. During the subsequent question and answer period, however, the President discretely moved toward the rear of the auditorium. Yet, as he approached the door, it was evident from the podium that the first slide of the second talk had captured his attention—the talk began with "This electron micrograph shows the fine structure of the heart and in particular...." The President remained standing at the door and listened to the entire 10-min talk. It is very important to capture the audience's attention quickly.

Finishing well is equally important to effective presentations. The conclusion is often that which the audience remembers best. Although Carnegie (1956) wrote on public speaking in general, not technical communication, it is interesting nonetheless to consider his suggestions for ending a talk: "summarizing, restating, outlining briefly the main points you have covered; appealing for action; paying the audience a sincere compliment; raising a laugh; quoting a fitting verse of poetry; using a biblical quotation; building up to a climax." Regardless of approach, ensure consistency between the opening and closing and try to memorize the ending so that it is thoughtful and forceful. Remember, too, that two of the best words to end with are "thank you."

It is important to embrace the question and answer period. Although some speakers prefer to avoid criticism and do not want to be questioned, one can obtain valuable suggestions and guidance during this exchange. Indeed, many times, one will learn something that will improve the quality of a subsequent paper that will be written on the topic of the presentation. Three useful guidelines are: first, repeat the question both to ensure that you address what was really asked and to help the audience to hear both question and answer; second, be respectful even if the questioner is antagonistic or if the question is truly a "dumb" question; and third, if you do not know the answer to the question, say that you do not know. It is best, however, not to answer all questions by stating that you do not know, hence the need for complete command of your subject. Finally, if a questioner tends to be unrelenting, suggest that you would enjoy discussing the issue at the next break. Remember, too, that because the question and answer period can be illuminating to both the speaker and the audience, finish the presentation early to allow sufficient time for this important exchange.

Although we addressed only the typical 15-min talk, presentations of other durations should be treated similarly. Indeed, if you become proficient at "telling your story" concisely, it is easy to do so for any specified duration. The one caveat, however, is to remember that you should always present concisely—a longer duration simply means that you should communicate more information, not that you should communicate the same information less well. Remember, too, that it is always good to think ahead about which slides to skip if time is running out or if there is an unavoidable delay or slowdown. In other words, be prepared and be flexible.

Exercise 5.7
Prepare a 30-min technical presentation, on a topic of your choice, using approximately 30 slides. After having given the presentation to your peers, reduce the presentation to 15 min without losing any significant technical content.

5.3   Internet-Based Presentations

Prior to March 2020, few of us had attended virtual conferences, seminars, lab meetings, or classes. Soon thereafter, however, nearly all of us had participated in many virtual meetings and events, most of which were surprisingly effective. In hindsight, it is remarkable how quickly we adjusted to this new method of communicating, although some did so more effectively than others.

The smooth transition to internet-based presentations stemmed in large part from the frequent use of a familiar tool, PowerPoint. Hence, most of the suggestions found in

Sect. 5.2 hold for internet-based presentations—that one should organize the presentation logically, minimize the amount of text on each slide via the ample use of schematic drawings, flow-charts, figures, and images, and explain in detail that which is on each slide. Toward this end, one should not try to present too much information and one should not rush. Using the ppt-provided "laser-pointer" was again very effective if the virtual laser was not moved excessively to the point of distraction.

Perhaps most difficult in this transition was the lack of interpersonal interactions, including loss of eye contact. Indeed, many of us realized for the first time just how influential nonverbal communication is for both presenter and participant. Effective presenters quickly transitioned from focusing on the audience, and facial expressions therein, to focusing solely on the material that was being presented. In some ways, it was best to imagine that you were speaking to a single person, one who was very interested in the topic. Indeed, when possible, it is best to have one or more listeners to keep their webcam active so that the presenter does not feel alone.

Another unexpected outcome of remote conferences was the need to pre-record some presentations so that they could either be managed effectively by the conference organizers or be played even if the presenter lost internet service. Being able to review a pre-recorded presentation prior to submission revealed to most of us many things that we could do to improve our presentations, including changes in word usage, reductions in the number of slides or content per slide, and pace of presentation. Indeed, we are all well-advised now to pre-record, preview, and revise presentations even if they are to be delivered in person.

Exercise 5.8
Use the ppt presentation that you designed in Exercise 5.7 and record your presentation using an internet-based platform such as Zoom. Review and revise the presentation and write a two-page summary of what you learned by reviewing your own presentation.

Exercise 5.9
Read and provide a three-page report on the suggestions for effective presentations given in: Naegle KM (2021) Ten simple rules for effective presentation slides. *PLoS Comput Biol* 17: e1009554.

5.4   The Technical Poster Presentation

Poster presentations have long been an important component of technical meetings and conferences. Although some investigators are disappointed to learn that their submitted abstract was accepted for a poster rather than a podium presentation, this time-honored method of presentation actually has many advantages. Most importantly, a poster presentation facilitates inter-personal interactions, literally conversations that may not otherwise be possible at small or large conferences.

Effective posters, like effective oral presentations, require careful planning and they improve with peer review prior to finalizing the poster. The challenge of designing an effective poster is actually greater than that for designing an effective oral presentation. A poster needs to achieve two parallel goals—to provide a clear and concise presentation to those who visit the poster in your absence and to provide visual aids for you as you explain your research in person. The former requires sufficient text to stand on its own, similar to a technical paper; the latter requires clear visuals to which you can refer for emphasis, similar to an oral presentation. There is, therefore, a need for balance, particularly noting that many visitors to a poster often will try to read text even while the presenter is talking. One strategy to accomplish both of these seemingly disparate goals is illustrated below.

Nearly all posters show the title and authors in large lettering along the top of the poster to attract attention. The title should thus be concise and engaging. The key question, however, is how best to distribute the remaining space? In this example, the main body of the poster is divided into three columns—narrow columns to the right and left and a dominant wider column in the middle. The narrow columns provide room for text that can be read in your absence. In contrast to a technical paper, however, one should note concisely the primary objectives (or, Specific Aims) rather than provide a scholarly introduction and one should note concisely the primary findings (or, Conclusions) rather than provide a comprehensive discussion. Both of these sections can thus benefit from brief introductory text followed by bulleted aims or findings. Written in this way, information can be assimilated quickly either in your absence or presence (as you point to specific bullets). Methods and Results can be written more in narrative form for those who visit the poster in your absence, remembering nonetheless that conciseness is key and font size must be sufficient for viewing the poster from 3 to 5 feet away. Most importantly, however, the central column can be reserved primarily to highlight critical methods (flowcharts, work-flows, schematic drawings, pictures) and primary results (figures preferred, but simple tables and equations as well). These materials need to have descriptive legends for those who visit the poster in your absence, but the font size should be smaller than that used for the actual visuals—the figure axes, tabulated entries, and so forth. Designed in this way, you can stand largely in front of either of the narrow columns (depending on whether you use your left or right hand to direct the conversation) and focus the attention of the listener(s) on the visuals within center of the poster while minimizing their temptation to read the legends while you are speaking. As you describe these results, it is best to move from top to bottom or from left to right, or both, as you refer to the visuals in order. Finally, it is appropriate to include acknowledgments (particularly funding sources), a few key references, disclosures, and of course contact information. These additional materials can be placed either in the lower right of the poster or along the bottom edge as illustrated here; they can be in a smaller font since they are provided mainly for completeness, not to draw attention. In addition to providing contact information, it can be useful to place business cards on one side of the poster for those who may be interested in contacting you. Some presenters even provide hardcopy printouts of the poster or a related preprint or paper.

Concise and Engaging Title Authors and Affiliations		
Specific Aims Brief intro, then • Aim 1 • Aim 2 • Aim 3 Brief summary of significance and innovation. **Methods** (text)	**Key Methods & Results** (visuals with descriptive legends) Schematic Drawings Images Figures Modest Tables Equations	**Results** (text) **Conclusions** Brief intro, then • Finding 1 • Finding 2 • Finding 3
Acknowledgements, References, Disclosures, Contact Information		

Whereas your first goal is to attract viewers to your poster, your second goal is to engage them by emphasizing the significance, innovation, and importance of the findings. It is critical, therefore, to know your material well and to present it confidently. One way to increase confidence is to practice presenting the poster in different ways knowing that different people will ask different questions to initiate the conversation. Remember, too, that a key advantage of a poster presentation is the inter-personal interactions. Hence, when someone approaches you to learn more about your work, first introduce yourself and ask them who they are. Next, thank them for coming to your poster and ask if they have specific questions. If yes, you can address their interests rather than present the material from beginning to end in a prepared, set manner. If, however, they say that they would like to hear more about your work, then you can direct them to the Specific Aims or overall objectives that are bulleted on the top left of the poster. Quickly redirect their attention to the middle of the poster, however, which highlights the primary findings. Finally, conclude by enumerating the important findings that are bulleted on the bottom right of the poster. Remember, too, that although one person may have initiated the conversation, many others may be drawn in and thus you should make eye contact with each person as they listen and ask if anyone else has questions.

In summary, an effective poster should attract interest and accomplish two aims: provide sufficient information to be appreciated in your absence and help you to communicate effectively when you are present. These multiple objectives are best achieved with a simple and visually appealing design that presents the material clearly and concisely. As we have emphasized throughout this book, we encourage the reader to develop an individual style that is effective and yet most comfortable. Toward this end, the poster design illustrated above is but one example; the possibilities for designs are endless. Most importantly, the final product should represent you, your co-authors, and your organization well while being sensitive to the different styles that might engage different people.

Exercise 5.10

Use the ppt presentation that you designed in Exercise 5.7 and build an associated poster that conveys the same information. Present the poster to peers and, based on their suggestions, consider revising the poster to increase clarity. Write a two-page summary of what you learned by creating and revising the poster.

Exercise 5.11

At the next technical meeting that you attend, visit posters and take notes on both effective and ineffective methods of communication in the absence of the presenter. Next, engage at least three presenters and keep mental notes on what was effective versus ineffective in their presentations. Write a three-page summary of lessons learned with the goal of using these lessons to improve your next poster and its presentation.

5.5   The 2-min Lightning Presentation

The increasing use of posters at technical meetings has led to another new style of presentation—the 2-min lightning talk. That is, to help poster presenters advertise their work before a larger audience, many meetings now allow short oral presentations to highlight that which can be found in individual posters. These presentations typically impose a one-slide, two-minute rule and presenters typically form a line to one side of the podium and present in rapid succession. There is, therefore, a pressing need to embrace concepts that have been emphasized throughout this chapter—engage the audience early, be well prepared, ensure clarity and conciseness, and finish on time.

It is important to remember in this regard that the goal is to garner interest, not to present the primary finding. Toward this end, the only text needed is the title of the presentation, the authors and affiliations, and the poster number (in a large font). The technical information should primarily be visual—show an engaging image or figure. While referring to these visuals, state in a few sentences, the motivation, the methods, and the implications. Importantly, invite any interested person to the poster, noting the poster number and time that you will be at the poster.

> **Exercise 5.12**
> Design two slides for a 2-min presentation. Record the presentation and critique it, working to increase clarity and conciseness. Write a two-page report on lessons learned via self-critique and iteration.

5.6 Closing Comments

It has been said that "Everyone has but one story to tell, they merely tell it in different ways to different people." We emphasize again, therefore, that one must know the intended audience. Rather than trying to sound scholarly, it is most important to be clear and effective. Avoid jargon; define terms carefully; read faces in the audience to obtain a sense of their understanding and engagement.

Recall from Chap. 1 that individual differences can bring a freshness and vitality to a field; individual personalities can generate excitement and interest. Each person should develop a style that is most effective and natural for them. The guidelines presented in this chapter are simply that, guidelines. We encourage the reader to consider or try the ideas presented here, but more importantly, to pay close attention to styles and techniques used by different speakers in different settings. You will be well served to take note of what is most effective and what is most ineffective, and to adjust your style accordingly.

References

Boorstin, D. J. (1983). *The Discoverers*. Vintage Books.

Carnegie, D. (1956). *How to Develop Self-Confidence and Influence People by Public Speaking*. Simon & Schuster.

Valiela, I. (2001). *Doing Science: Design, Analysis, and Communication of Scientific Research*. Oxford University Press.

Authorship

6

Seeing your name on your first published paper may be one of the most exciting moments in your career. Many students thus enter a discussion on authorship focusing on the question, "How do I get my name on a paper?" In our experience, more important and difficult questions include when and how to keep your name off a particular paper and how to negotiate questions of authorship among collaborators in a multi-investigator project.

Exercise 6.1
If you have authored a journal article, answer the following questions about your experience before proceeding. If not, interview someone who has authored a journal article and report their answers.

1. How did you become an author on your first paper?
2. What was your contribution to that paper?
3. Who decided whose names would appear and in what order?
4. At what point during the research did you first discuss authorship?
5. Did you sign a legal agreement as an author, and if so, to what did you agree?

6.1 The Slutsky Case

Many widely publicized cases of plagiarism, research fraud, and other forms of misconduct exist in science and engineering. Discussing these cases often sheds light on important aspects of ethics in science and engineering. We take as an example the case of Dr. Robert Slutsky, a member of the faculty of the School of Medicine at the University

© The Author(s), under exclusive license to Springer Nature Switzerland AG 2024
J. D. Humphrey and J. W. Holmes, *Style and Ethics of Communication in Science and Engineering*, Synthesis Lectures on Engineering, Science, and Technology,
https://doi.org/10.1007/978-3-031-39125-5_6

of California San Diego in the 1980s. While in many ways similar to other cases of plagiarism or data fabrication, the Slutsky case is unusual because the university committee formed to investigate allegations of research fraud against Dr. Slutsky included a philosopher in addition to medical school faculty, and the committee attempted to draw broader conclusions about this type of misconduct. The committee published its findings in an article in *The New England Journal of Medicine* (Engler et al., 1987), and its investigation served as a template for many later investigations (Sox & Rennie, 2006). We briefly review details of the case below, but this excellent article is so integral to our discussion that it should be read before proceeding.

Exercise 6.2

Read the journal article regarding the Slutsky case [Engler RL, Covell JW, Friedman PJ, Kitcher PS, Peters RM (1987) Misrepresentation and responsibility in medical research. *N Engl J Med* 317: 1383–1389] and list the three aspects of the case that you find to be most surprising:

1.

2.

3.

Dr. Slutsky, then Associate Professor of Radiology at the University of California San Diego, was being evaluated for tenure when a member of the tenure committee noticed an apparent duplication of data in two published research papers. The ensuing investigation by a faculty committee revealed a number of striking facts of interest for our discussion. First, the committee found clear evidence that Dr. Slutsky reported fictitious experiments and statistical analyses, reported incorrect procedures and statistical analyses, and listed colleagues as coauthors who did not contribute to the work and in some cases did not know about the publications. Second, the normal peer review process did not detect any of these concerns in the fraudulent papers, and some of the journals refused to retract the fraudulent papers upon notification of the committee's findings unless Dr. Slutsky agreed to the retractions. Third, at one point during the period under investigation, Dr. Slutsky was publishing a paper every 10 days, including many in prestigious journals. Fourth, much of this work was apparently sound; the committee established the validity of 77 of the 137 publications they reviewed and classified only 12 publications as fraudulent. Finally, the investigation revealed missed warning signs over the course of Dr. Slutsky's early career: several of his colleagues and at least one journal editor questioned the validity of manuscripts they read, and some recommendation letters for his original appointment to the faculty expressed concerns about the validity or quality of his research.

6.2   Basic Conventions

Before discussing common problems regarding authorship, it is helpful to review current conventions. These conventions will be familiar to practicing scientists and engineers but not necessarily to undergraduate and graduate students, particularly those who have not yet authored a journal paper.

6.2.1   Order of Authors

The order of authors on an archival journal paper usually has special significance, but conventions vary by field and occasionally by journal. In most biomedical science and engineering journals, the first author is usually the one who performed most of the work; this person is often a graduate student or postdoctoral fellow who worked on the project described in the publication. Designation as first author is so important that footnotes are sometimes used to indicate equal contribution by two or more "first" authors. The last author is typically the senior investigator who conceived, guided, and financially supported the project. The ordering of the other authors is generally of less significance, as we assume that their contributions were less but otherwise important. In stark contrast, some fields encourage an alphabetical listing of authors. Notwithstanding customary variations by field and journal, surveys of scientists and engineers have traditionally revealed widespread disagreement and confusion regarding conventions for authorship (Bhopal et al., 1997; Tarnow, 1999).

6.2.2   Submission Agreement

Most journals ask the author(s) to sign a submission agreement. Typically, this agreement asks the author(s) to verify the accuracy of the submitted manuscript and that it has not been submitted to or published by another journal. Much more variable are policies regarding who must sign this and related agreements. In many cases, the corresponding author (i.e., the person who submits the manuscript and lists his or her contact information in the final version) can sign on behalf of all coauthors. This practice explains, in part, how Dr. Slutsky submitted some papers without the knowledge of some of the people he listed as coauthors. Conversely, some journals require all coauthors to sign the submission agreement; it appears that Dr. Slutsky subverted this requirement by forging the signature of some coauthors (Engler et al., 1987). Such forgery is less likely today with on-line verifications. Finally, either this agreement or another one at the time of acceptance requires transfer of copyright to the publisher and assurance that the author(s) will pay charges levied by the journal as part of the publication process.

6.2.3   Publication Impact

One's record of publication is critically important when applying for jobs, grants, awards, or promotion and tenure. In any discussion of authorship, it helps to understand how reviewers evaluate your published works. Obviously, the number of publications is an important factor, but this is far from the only consideration. For example, some journals are more selective and more widely read than others; publications in these journals are often valued more in an evaluation. Such assessments are subjective, however, because investigators in the same field may have different opinions on the relative quality of the relevant journals or their ability to assess the quality of a particular work. For example, a complex mathematical model of a biological process will likely receive a more rigorous review by a mathematics journal than by a biology journal even though the latter may have a larger readership. In an attempt to weigh the quality of a journal more objectively, one can define quantitative metrics. One such metric is the Journal Impact Factor, a measure based on the idea that more frequent citations of a journal's articles implies a greater impact by that journal on its field. The impact factor is computed and published in Journal Citation Reports by the company Clarivate; other similar journal metrics include CiteScore by Elsevier/Scopus and the SCImago Journal Rank. Although well-known and commonly used, there remains considerable debate regarding the utility of such metrics. As an example, these citation records do not assess whether a given paper was cited positively or negatively; given the need to motivate papers and grant proposals, one must necessarily cite papers to demonstrate prior limitations.

Even if one relies on a metric such as impact factor to value a publication, most publications have multiple authors. The question then becomes, "How much credit should each author receive for a given publication?" For example, consistent with conventions discussed above regarding the order of authors, the first and last authors typically receive most of the credit for any biomedical publication. When someone evaluates your publication record, they will notice not only the number of publications and the quality of the journals but also how often you appeared first or last in the author list. Note, however, that even if your name appears last on a publication, implying that you were the senior author most responsible for the ideas, if a well-recognized senior colleague also appears on the paper, other scientists may assume that your senior colleague deserves much of the credit for the ideas. This issue becomes especially relevant in multi-investigator collaborations involving a mix of junior and senior scientists.

6.3   Common Problems

In June 2005, an article in the journal *Nature*, titled "Scientists Behaving Badly," reported results from a survey of more than 3000 NIH-funded scientists regarding the frequency with which they engaged in a range of questionable research practices (Martinson et al.,

2005). While only 0.3% admitted to falsifying research data within the previous 3 years, many more admitted to some of the other problems highlighted by the Slutsky case, such as publishing the same data in two or more publications (4.7%), or inappropriate assignment of authorship within the past 3 years (10%). The *Nature* article inspired a wide range of other studies that explored self-reported or observed questionable research practices in specific disciplines (Eisenberg et al., 2011; Pruschak & Hopp, 2022; Rajasekaran et al., 2014; Wislar et al., 2011), in specific countries or regions (Godecharle et al., 2018; Gopalakrishna et al., 2022; Tijdink et al., 2014), or in the biomedical industry compared to academia (Godecharle et al., 2018). Some of the most interesting studies examined the relationship between questionable research practices and the pressure or degree of competition experienced by researchers (Fanelli, 2010; Gopalakrishna et al., 2022; Necker, 2014; Tijdink et al., 2014).

As a group, these studies found surprisingly high rates of inappropriate authorship in the biomedical literature. Wislar and colleagues found that the corresponding author of 29% of publications in top medical journals in 1996 and 21% of publications in those same journals in 2008 believed that at least one author on their paper did not merit authorship (Wislar et al., 2011). Discipline-specific surveys found rates of self-reported inappropriate authorship of 15–30% (Eisenberg et al., 2011; Pruschak & Hopp, 2022; Rajasekaran et al., 2014). Not surprisingly, these self-reported rates may underestimate the actual scope of the problem: more than 50% of the same survey respondents agreed that at least one author on their paper performed tasks that would not qualify for authorship under International Committee of Medical Journal Editors (ICMJE) standards (Eisenberg et al., 2011; Rajasekaran et al., 2014). Furthermore, surveys that ask scientists questions about practices they have observed among colleagues produce truly shocking results: the vast majority (72–85%) of respondents report observing others adding authors who did not contribute substantively to the study or manuscript (Godecharle et al., 2018; Tijdink et al., 2014). A particularly interesting paper by Pruschak and Hopp had respondents answer questions not only about the tasks performed by authors on their papers but also about hypothetical scenarios. Their results suggest that many scientists apply less stringent criteria when determining authorship than recommended by journals via the ICMJE, particularly when students or postdoctoral fellows are involved (Pruschak & Hopp, 2022); the ICMJE standards are discussed further in Sect. 6.4.

6.3.1 Expectations

The importance placed on publications as a measure of career progress can create substantial pressure to publish, particularly for tenure-track junior faculty. Managing this pressure begins by developing clear and reasonable expectations.

Exercise 6.3

For Ph.D. students and postdoctoral fellows, answer the following questions regarding the number of publications you expect a junior faculty member to produce in your field. First, estimate the number of publications one might produce (or that you produced) during doctoral study. Next, estimate the number of publications one might produce (or that you produced) during 3 years of postdoctoral research. Finally, estimate the number of publications successful junior faculty members in your field should produce during the first 5 years of their career. Now, perform two different "reality checks" on your estimates:

1. Translate your estimates of productivity into rates (papers per year, which may be < 1), noting that most papers tend to be produced near the end, not beginning, of one's study. Then, use these rates to compute how many graduate students and/or postdoctoral fellows your hypothetical junior faculty member would need to employ if each paper was coauthored by only one student or fellow. Do you think these numbers are reasonable? Which of your estimates would you adjust based on this check?
2. As a second check, ask a senior faculty member in your department to give the same three estimates. How do they compare to your estimates? If possible, discuss any discrepancies with that faculty member or your mentor.

One thing that seems apparent in the case of Dr. Slutsky is an unrealistic expectation or perception of expectations regarding productivity. No reasonable person expects a junior faculty member in any field to produce a paper every 10 days. Yet, Dr. Slutsky apparently felt pressure to improve upon his already large number of valid publications (at least 77 in 7 years according to the authors of the report in the *New England Journal of Medicine*) through various types of research fraud. Clearly, there is a need for open discussion of authorship and productivity with everyone involved, from students to advisers to department chairs, to develop and communicate realistic expectations regarding the number and quality of publications.

6.3.2 Gift, Guest, and Ghost Authorship

Gift authorship entails granting authorship to a person who did not contribute directly to the work (Davidoff, 2000). As an example, a new trainee discovered upon her arrival in Dr. Slutsky's laboratory that she was already an author on a paper she knew nothing about (Engler et al., 1987). Why would someone do this? Misplaced generosity could be one motivation—colleagues may believe they are doing you a favor by listing you as an author on a publication. Another possible reason could be pressure to show productivity by trainees who are supported by certain types of grants. Regardless, gift authorship is wrong and in extreme cases it could associate your name with a fraudulent paper, as in the Slutsky case. Cases of research fraud are rare, however; embarrassment is a more likely concern if you consider the paper to be of poor quality, you disagree with its conclusions, or you are compelled to admit (e.g., during the question and answer period after a scientific talk or during a discussion with a colleague) that you did not contribute to the work.

It is easy to prevent gift authorship when offered. By contrast, one of the most difficult situations related to authorship is learning of an unwanted gift authorship after publication, especially if you are a junior colleague of the person conferring it. What choices did the trainee in Dr. Slutsky's laboratory have when she was told about the gift authorship? What would you have done? Two points are appropriate here. First, there exists an anonymous procedure at most universities and companies for seeking advice if you encounter a situation such as gift authorship; typically, an officially designated ombudsman will help you to resolve conflicts and difficult situations. Second, an adviser who puts you in a precarious situation is likely not the right adviser for you. The consequences of confronting a situation like this early are likely not as bad as they seem, while the consequences of avoiding confrontation are likely much worse.

Studies of scientific authorship often define guest authorship separately as listing a colleague who did not contribute directly to a paper in the hope that his or her reputation will enhance the odds of acceptance for publication (Davidoff, 2000). Another category that falls under the umbrella of inappropriately assigning credit is ghost authorship, defined as the omission of an author who contributed significantly to the publication. Ghost authors may be junior colleagues who simply did not receive the credit they deserved, but they may also be professional medical writers hired to write articles anonymously or even representatives from companies with a financial interest in the findings who wish to hide their involvement.

While trainees may feel that they have little input into decisions of authorship, their first experiences with authorship may shape their attitudes and future practices. In one particularly relevant study, 38% of postdoctoral research fellows at the University of California San Francisco reported that at least one coauthor on their papers was undeserving, while 20% thought they were excluded on at least one paper for which they deserved authorship (Eastwood et al., 1996). Overall, 32% of the fellows surveyed said they would be willing to list an undeserving author on a paper if it would enhance the probability of publication or otherwise benefit their career, but that number jumped to 72% among those who reported a previous adverse experience with authorship.

6.3.3   Financial Support

Accepted practice regarding authorship varies by field and by culture. While relatively rare, some strongly hierarchical departments expect the chair of the department to be listed as an author (possibly even senior author) on every paper, regardless of contribution. Such an environment may pose a challenge for younger investigators who disagree with the policy, especially if following the expected procedure weakens their publication records by preventing them from assuming senior authorship on their work. It is important to recognize and discuss cultural variations when working in a group composed of colleagues who trained under different systems and when collaborating internationally.

6.3.4   Quid Pro Quo

Nearly everyone agrees that gift authorship is wrong, yet there are many related cases where a colleague who has contributed to a study in some way requests or expects authorship in return. The most common situation involves valuable resources such as antibodies or transgenic mice. Consider a situation where an investigator devotes significant time and energy to develop such a resource, then publishes a paper describing it. Colleagues read the paper and ask for access to the resource for their own studies. It is not uncommon for the investigator to offer to provide access in exchange for authorship on the resulting paper(s).

This basic situation has unlimited variations. At one extreme, a request for a resource can lead to a genuine collaboration on a new study that is reflected accurately in coauthored papers describing the results. At the other extreme, however, the situation can approach scientific extortion, with the original developer of the resource demanding authorship in exchange for access, knowing few colleagues will deny the request due to the substantial time and effort required to recreate the resource. While many who disagree with such arrangements accept them as a fact of life, some defend the practice, regarding authorship on related papers as appropriate reward for originating the resource.

Most believe that the appropriate reward for any innovation, whether a new equation, method, antibody, or transgenic mouse, is citation, not authorship. Colleagues who employ the innovation cite the original publication, giving appropriate credit to its originator. In the case of an equation or its solution, the original paper contains everything colleagues need. In the case of a transgenic mouse, the original paper contains only a description of how to generate such a mouse. Is it reasonable to expect the scientist who first generates the mouse to send mice to any colleague who requests them? Does the answer change if federal or state resources funded the original development of the mouse, as with most biomedical research? These and related questions about access to resources and data from publicly funded science have long generated vigorous discussion in the scientific community. Current trends towards open access and transparency further render this discussion important and timely.

6.3.5 Students and Technicians

We have highlighted some common problems related to authorship beginning with the simplest and least controversial and proceeding to the more complex and controversial (and therefore interesting). Next, we consider the key question of who should or should not be an author on a particular paper or, to generalize the problem, of exactly what qualifies someone to be an author. Before proceeding, use the following exercise to define better what you think should be considered in making decisions about authorship.

> **Exercise 6.4**
> List up to five minimum criteria needed to justify authorship on a scientific or engineering paper. According to your criteria, would a laboratory technician or an undergraduate student who orders supplies and prepares samples qualify as an author on papers produced by the laboratory? What about a technician or student who runs tests according to instructions and turns over the data for analysis? What about one who runs tests, analyzes data, and makes a figure for the paper but does not write any of the text?
> 1.
> 2.
> 3.
> 4.
> 5.

As this exercise illustrates, it can be difficult to articulate general guidelines for authorship that provide practical guidance. Common responses to this exercise are that each author should make a "significant" contribution to the work, that each author should make an "essential" contribution to the work, or that each author should make an "intellectual" contribution to the work. This last point illustrates general agreement that a student or technician who simply prepares samples or collects data without a true understanding of the project should be acknowledged, not listed as an author. Nevertheless, none of these statements provides practical guidance.

To increase our appreciation of this situation, it is useful to consider the contribution of a potential author against the backdrop of what is required to produce a paper. First, one must generate an idea or identify a problem, then plan an approach to address the need. Next, one must perform the study and collect the data or solve the equations. Analysis and interpretation of the data or results precedes writing the paper, which typically requires a comparison to previous related findings and thus broad knowledge of the archival literature. The example of a student or technician who only collects data or runs a computer code as instructed suggests that an author should be involved in more than one aspect of the study; if that person also analyzes data and summarizes the results for the paper, the claim to authorship would be stronger. Requiring involvement in multiple aspects of a study would limit the quid pro quo arrangements discussed above to cases where involvement went beyond providing a particular resource. It seems reasonable to stop short of requiring every author to participate in every phase, however, particularly as scientific and engineering investigations become more complex and multidisciplinary. For example, many investigators would support authorship for a person who joined a group after the study was conceived and planned, yet participated fully in all aspects of the actual work.

The concept that all authors should be involved in multiple aspects of a study (e.g., design, experiment, analysis, interpretation, or writing) seems reasonable. Nevertheless, your list from Exercise 6.4 likely includes additional criteria. Must every author understand everything in the paper? Must every author read the final version before submission? Recalling that some of Dr. Slutsky's coauthors experienced the stigma of being authors on fraudulent papers, should every author review the original data that form the basis for the conclusions? Each investigator must wrestle with these questions over the course of a career; your answers to these questions may well evolve with experience. It is important to think carefully about these issues early in your career so that you can develop practices consistent with the ethical standards you set for yourself.

6.4 Current Standards and Ongoing Debates

Many people have thought about ways to define authorship in the archival literature. In particular, some professional societies and journals have introduced simple practices that reflect more accurately the contributions of those involved in a publication. These practices also have the beneficial effect of forcing increased discussion among coauthors on issues related to authorship.

6.4.1 International Committee of Medical Journal Editors Standards

The International Committee of Medical Journal Editors (ICMJE) evolved from meetings that began in 1978 to establish guidelines for the format of manuscripts submitted to medical journals. Their early publication, "Uniform Requirements for Manuscripts Submitted to Biomedical Journals," focused primarily on conventions for organizing manuscripts into sections, standardizing reference formats, and so forth. Over the years, ICMJE grew to encompass a much broader range of publication-related questions including those of authorship, fraud, retractions, and other topics. Today, www.ICMJE.org provides guidelines under the title, "Recommendations for the Conduct, Reporting, Editing, and Publication of Scholarly work in Medical Journals," as well as a list of journals that employ ICMJE standards. This brief summary of recommendations is a valuable resource and should be read.

As noted in previous sections, many studies show that scientists appear to assign authorship on their publications using less stringent standards than advocated by the ICMJE. In addition, changes in the ICMJE criteria themselves over the years reflect the ongoing debate in the scientific community about authorship. The earliest guidance from ICMJE stated simply, "Acknowledge only persons who have made substantive contributions to the study" (Huth & Case, 2004). By 1998, the guidelines included the statement, "Each author should have participated sufficiently in the work to take responsibility for the content." This statement prompted robust debate. In a large collaborative study involving dozens or even hundreds of co-authors, is it realistic to expect that every co-author can vouch for the integrity of the work performed by every other team member? To address this situation, ICMJE introduced the idea of one or more team members who would take final responsibility for a group effort, and restricted the scope of their statement on responsibility. In 2004, the guidelines stated (ICMJE 2004):

> Authorship credit should be based on: (1) Substantial contributions to conception and design or acquisition of data or analysis and interpretation of data; (2) Drafting the article or revising it critically for important intellectual content.; and (3) final approval of the version to be published.

> When a large, multi-center group has conducted the work, the group should identify the individuals who accept direct responsibility for the manuscript...

> Acquisition of funding, collection of data, or general supervision of the research group, alone, does not justify authorship.

> All persons designated as authors should qualify for authorship, and all those who qualify should be listed.

> Each author should have participated sufficiently in the work to take public responsibility for appropriate portions of the content.

How do these criteria compare to your answers from Exercise 6.4 above? Many trainees spontaneously produce a list similar to the first three numbered criteria from the 2004 ICMJE guidelines, and those statements have remained a stable component of the guidelines for more than 20 years. Nevertheless, discussions and guidelines on how to address accountability in cases of fraud or error continue to evolve. In 2022, the guidelines stated (ICMJE 2022):

> The ICMJE recommends that authorship be based on the following 4 criteria:

1. Substantial contributions to the conception or design of the work; or the acquisition, analysis, or interpretation of data for the work; AND
2. Drafting the work or revising it critically for important intellectual content; AND
3. Final approval of the version to be published; AND
4. Agreement to be accountable for all aspects of the work in ensuring that questions related to the accuracy or integrity of any part of the work are appropriately investigated and resolved.

In addition to being accountable for the parts of the work he or she has done, an author should be able to identify which co-authors are responsible for specific other parts of the work. In addition, authors should have confidence in the integrity of the contributions of their co-authors.

6.4.2 Author Notification

Many conferences and journals now require the corresponding author to provide e-mail addresses for all authors, who are notified electronically of the submission of an abstract or manuscript. While no coauthor should ever learn of a submission for the first time through such an e-mail, this is not an infrequent occurrence. Notification allows an investigator who was unaware of a submission to raise objections while the abstract or manuscript is under review, rather than being forced into the much more difficult position of addressing the issue after acceptance or publication. Notification may also increase the odds that the corresponding author will discuss the submission with all coauthors in advance to avoid surprising colleagues. Electronic notification is not a foolproof defense against those who are willing to forge the names of coauthors on a submission agreement. Those intending to deceive could easily construct false e-mail accounts for coauthors, but at least this would require more effort than simply forging a signature.

6.4.3 Specifying Contributions

In concert with the discussion on how to apportion responsibility and accountability among co-authors of a collaborative study, many journals began asking authors to specify their contributions to an article at the time of submission. Specifying individual contributions simplifies attribution of responsibility or blame. It could also allow societies or journals to impose more uniform standards for authorship. For example, a journal could refuse authorship to anyone unwilling to take responsibility for more than one aspect of a publication. Partitioning responsibility may be the only practical solution for large multi-investigator projects. Nevertheless, this approach changes the traditional understanding of an archival publication and meaning of authorship. It could have the disadvantage of weakening scientific collaborations if journal papers become simply a compendium of individual miniprojects. Such a weakening is certainly contrary to what most of us envision when we discuss the need to foster more and better multidisciplinary collaboration on today's increasingly complex scientific and engineering problems.

6.4.4   Quantifying Contributions

A natural response to uncertainty, especially among scientists and engineers, is to introduce quantitative measures. In addition to specifying what each author did, some have advocated specifying each author's percent contribution to the overall work. This is probably most common during tenure evaluation, when a junior faculty member under consideration for tenure estimates his or her percent contribution to each published paper. This is a difficult question to answer, especially during a tenure evaluation, because the desire to report strong contributions for yourself may tempt you to devalue the contributions by your coauthors. Typically, a group of collaborators who estimate the contribution of each group member produce percentages greater than 100% unless some mechanism (such as an interactive form or pie chart) constrains the total.

Another quantitative approach appeared in the biostatistics literature, reflecting the unique role played by many statisticians in research. Statisticians may be involved in design and analysis for many different studies but not directly involved in collecting data, performing experiments, or writing papers for any of those studies. This "specialist" role makes it difficult to apply typical criteria for authorship. As a possible solution to this problem, Parker and Berman (1998) proposed a scoring system to help decide when statisticians should or should not be listed as authors. Their system divides the statistician's role into three phases of a research project (design, implementation, and analysis) and requires for authorship either a deep involvement in two of the phases or a deep involvement in one and moderate in the other two. They also proposed that it is unreasonable to hold a statistician who is listed as an author responsible for the integrity of the entire published article.

6.5   Our Approach

As is common when discussing interesting ethical issues, we raised many more questions than we answered in this chapter. What is most important is that each person utilizes cases and questions such as those presented herein, as well as discussions with advisers and senior colleagues, to establish individual principles about authorship early in a career. *It is impossible to do what you think is right if you do not know what you think is right.* Once you establish your principles, the question remains of how best to put them into practice. In this section, we offer some of our own experiences as examples of how to apply a set of principles to the everyday practice of science and engineering.

6.5.1 Authorship Criteria

In our own groups, we expect that all authors on a paper should be involved in more than one aspect of a study, should agree to be listed as an author, and should be given a chance to contribute directly to the final version of the manuscript before submission. Ideally, each new group member and each new collaborator should discuss these criteria at the outset. At the very least, all members of the group working on a particular project must discuss issues of authorship before submitting the first abstract or publication related to that work. This approach has typically been easiest to accomplish when all authors work at a single location, but the advent of convenient internet-based web conferencing has rendered this step equally tractable for colleagues from different departments or institutions.

6.5.2 Predraft Group Meeting

In our experience, one of the simplest and most useful ideas is to convene a meeting of all potential authors to review findings and interpretations as well as to agree on authorship before writing an abstract or manuscript. The senior investigator who is funding or driving the project calls the meeting, inviting all contributors who potentially satisfy the criteria for authorship. In cases of coauthors from multiple locations, web conferencing becomes a vital resource. At the meeting, each contributor presents results to the group and answers questions. Then, the group discusses proposed figures, the proposed author list, and the choice of journal for submission. Notwithstanding the effort required to bring everyone together for an hour or two, this approach allows all potential authors to gain confidence in the validity of the studies, to ask questions and comment on the results and their importance, and to voice any concerns about the content of the paper, interpretation of the results, or author list before the bulk of the writing begins. This approach also helps to improve the paper by subjecting the results to a round of "internal review," helps graduate students and fellows practice oral presentation skills, and helps strengthen relationships among collaborators.

6.5.3 Final Review and Approval

Once a manuscript becomes a final draft, it is essential for all authors to review and approve the draft before submission. This is also an appropriate time to settle final questions of authorship, especially if no previous discussion has taken place. One reasonable approach is to list as authors on the draft those colleagues you believe merit authorship, but to include in the distribution list other people who have made some contribution and may feel that they should be authors. Ask each recipient whether they feel they deserve to be an author on the paper (or whether they agree with the proposed author list) and whether they have any comments or suggestions for the manuscript before submission; follow up with those who do not reply. Like the predraft group meetings, this step ensures that all authors are aware of the content of any publication bearing their names and provides a round of internal review to improve the manuscript before external peer review.

Most investigators basically agree on the rules of authorship and are willing to follow them. Inappropriate attribution of authorship usually reflects someone succumbing to real or perceived external pressures or simply not giving the matter sufficient attention rather than attempting to deceive. In general, our experience with regard to questions of authorship has been heartening. In most cases where a claim to authorship appeared marginal to us, our colleagues have responded to our question of whether they want to be an author, as we would have hoped, by stating that their contribution merits an acknowledgment rather than authorship. Many have provided helpful comments on a draft even after stating that they did not wish to be listed as authors. Perhaps surprisingly, our most difficult experiences have involved refusing authorship offered by a colleague rather than denying authorship to a colleague.

6.5.4 Default Position for Abstracts

The process described above is time-intensive. A confounding situation that can arise, therefore, is the last-minute abstract for consideration for presentation at a technical meeting. Such abstracts are short and typically have a fixed deadline for submission, thus they are often written just before the deadline. On such short notice, the collaborators involved in a particular study may not be available to meet to discuss the abstract or even to read, revise, and approve the final submission. In such cases, it is best to encourage early preparation but at least to agree ahead of time on a "default" position for last-minute abstracts—if contributors cannot be reached to review an abstract on short notice, do they prefer to be listed as an author and review the abstract after submission or do they prefer to be left off the author list? We recommend the latter approach, for it is dangerous practice to include authors who have not read, revised, and approved the abstract before submission. We also note that it is appropriate to delay submission when a coauthor cannot be reached; there will always be other meetings and thus other opportunities.

References

Bhopal, R., Rankin, J., McColl, E., Thomas, L., Kaner, E., Stacy, R., Pearson, P., Vernon, B., & Rodgers, H. (1997). The vexed question of authorship: Views of researchers in a British medical faculty. *BMJ, 314*, 1009–1112.

Davidoff, F. (2000). For the CSE Task Force on Authorship. Who's the author? Problems with biomedical authorship, and some possible solutions. *Science Editor, 23*, 111–119.

Eastwood, S., Derish, P., Leash, E., & Ordway, S. (1996). Ethical issues in biomedical research: Perceptions and practices of postdoctoral research fellows responding to a survey. *Science and Engineering Ethics, 2*, 89–114. https://doi.org/10.1007/BF02639320

Eisenberg, R. L., Ngo, L., Boiselle, P. M., & Bankier, A. A. (2011). Honorary authorship in radiologic research articles: Assessment of frequency and associated factors. *Radiology, 259*(2), 479–486. https://doi.org/10.1148/radiol.11101500

Engler, R. L., Covell, J. W., Friedman, P. J., Kitcher, P. S., & Peters, R. M. (1987). Misrepresentation and responsibility in medical research. *New England Journal of Medicine, 317*, 1383–1389.

Fanelli, D. (2010). Do pressures to publish increase scientists' bias? An empirical support from US States data. *PLoS ONE, 5*(4), e10271. https://doi.org/10.1371/journal.pone.0010271

Godecharle, S., Fieuws, S., Nemery, B., Dierickx, K. (2018). Scientists still behaving badly? A survey within industry and universities. *Science and Engineering Ethics, 24*, 1697–1717. https://doi.org/10.1007/s11948-017-9957-4

Gopalakrishna, G., Ter Riet, G., Vink, G., Stoop, I., Wicherts, J. M., & Bouter, L. M. (2022). Prevalence of questionable research practices, research misconduct and their potential explanatory factors: A survey among academic researchers in The Netherlands. *PLoS ONE, 17*(2), e0263023. https://doi.org/10.1371/journal.pone.0263023

Huth, E. J., & Case, K. (2004). The URM: Twenty-five years old. *Science Editor, 27*(1), 17–21. https://www.icmje.org/recommendations/archives/summary78-04.pdf.

Martinson, B. C., Anderson, M. S., & de Vries, R. (2005). Scientists behaving badly. *Nature, 435*, 737–738. https://doi.org/10.1038/435737a

Necker, S. (2014). Scientific misbehavior in economics. *Research Policy, 43*(10), 1747–1759. https://doi.org/10.1016/j.respol.2014.05.002

Parker, R. A., & Berman, N. G. (1998). Criteria for authorship for statisticians in medical papers. *Statistics in Medicine, 17*, 2289–2299. https://doi.org/10.1002/(SICI)1097-0258(19981030)17:20%3c2289::AID-SIM931%3e3.0.CO;2-L

Pruschak, G., & Hopp, C. (2022). And the credit goes to …—Ghost and honorary authorship among social scientists. *PLoS ONE, 17*(5), e0267312. https://doi.org/10.1371/journal.pone.0267312

Rajasekaran, S., Shan, R. L. P., & Finnoff, J. T. (2014). Honorary authorship: Frequency and associated factors in physical medicine and rehabilitation research articles. *Archives of Physical Medicine and Rehabilitation, 95*(3), 418–428. https://doi.org/10.1016/j.apmr.2013.09.024

Sox, H. C., & Rennie, D. (2006). Research misconduct, retraction, and cleansing the medical literature: Lessons from the Poehlman case. *Annals of Internal Medicine, 144*(8), 609–613. https://doi.org/10.7326/0003-4819-144-8-200604180-00123

Tarnow, E. (1999). The authorship list in science: Junior physicists' perceptions of who appears and why. *Science and Engineering Ethics, 5*, 73–88. https://doi.org/10.1007/s11948-999-0061-2

Tijdink, J. K., Verbeke, R., & Smulders, Y. M. (2014). Publication pressure and scientific misconduct in medical scientists. *Journal of Empirical Research on Human Research Ethics, 9*(5), 64–71. https://doi.org/10.1177/1556264614552421

Wislar, J. S., Flanagin, A., Fontanarosa, P. B., & Deangelis, C. D. (2011). Honorary and ghost authorship in high impact biomedical journals: A cross sectional survey. *BMJ, 343*, d6128. https://doi.org/10.1136/bmj.d6128

Recordkeeping, Reproducibility, and Responsibility

7

Scientists and engineers must keep records of their work, using a combination of note-books, images, and electronic documents and data. Similarly, medical professionals must record each step of diagnosis and treatment in a patient's medical records. Although keeping precise records may seem mundane, those records are central to many important decisions in science and engineering just as they are in law, medicine, and public policy.

Exercise 7.1

Based on your experience in a research laboratory, or a laboratory course if you have not yet worked in a research laboratory, answer the following questions before continuing:

1. Did you maintain a paper or electronic notebook?
2. If yes, what instructions were you given about what to record?
3. If no, where did you record information related to the work?
4. Did your supervisor review your notebook or records? If yes, how often?
5. If someone tried to reconstruct your work from these records, what percentage could they reconstruct without your help?

If possible, compare your answers to those of a colleague who has worked in the pharmaceutical, medical device, or software industry. It is likely that your answers will differ substantially; discuss the most likely reasons for this.

© The Author(s), under exclusive license to Springer Nature Switzerland AG 2024   121
J. D. Humphrey and J. W. Holmes, *Style and Ethics of Communication in Science and Engineering*, Synthesis Lectures on Engineering, Science, and Technology,
https://doi.org/10.1007/978-3-031-39125-5_7

7.1   The Slutsky Case Revisited

In Chap. 6, we considered the case of research fraud by Dr. Robert Slutsky, as described in a 1987 article in the *New England Journal of Medicine* (Engler et al., 1987), and we asked what aspects of this case were most surprising. In response to this question, many cite the following paragraph from the section entitled "What is Fraud?"

> After due consideration of what requirements and standards applied, the ... committee adopted the position that the ethos of scientific research requires that hypotheses be validated before they can be accepted and that claims to observation be open to scrutiny by peers. The legal principle of "innocent until proved guilty," which might be rephrased as "assume correct until proved wrong," does not apply to scientific work; the burden of proof remains with those claiming new findings. Thus, the authors of a scientific publication that is reasonably alleged to be fraudulent bear the responsibility for establishing the accuracy of their results.
>
> Engler et al. (1987)

This excerpt should be sobering to anyone involved in research or development. Most of us have lost records to a computer crash, accidentally overwritten a file, lost a notebook, discarded old data, or at times kept imperfect records. If the supporting data are missing and the burden of proof against an allegation of fraud lies with the researcher, an anonymous accusation of fraud from a disgruntled colleague, employee, or student could be enough to support a finding of research fraud and possibly end a career.

The proposition that the burden of proof lies with the researcher raises two questions. First, do you agree with the argument that the nature of science and engineering should place the burden of proof on the investigator, or should it rest (as in criminal law) with the accuser? Second, is the burden of proof actually placed on the researcher in current practice? While the first question provokes interesting discussions in any room of scientists or engineers, most are surprised to learn that the answer to the second question is a resounding *yes*.

The Office of Research Integrity (ORI) of the U.S. Department of Health and Human Services (http://ori.hhs.gov/) performs a range of functions designed to maintain the scientific integrity of biomedical and behavioral research funded by the U.S. Public Health Service (PHS). One of these functions is investigating and issuing reports on scientific misconduct involving PHS grants. Moreover, to heighten awareness of the importance of scientific integrity, the ORI publishes Findings of Scientific Misconduct (i.e., brief reports summarizing each case and its outcome). Recent cases are listed on their website; older cases are available in PubMed Central and the Federal Register. The exercise below intentionally uses an older example, but the standards it illustrates remain. Particularly relevant to our discussion is that in the case noted below, the ORI found that the researcher "engaged in scientific misconduct by reporting research that was inconsistent with original data or could not be supported because original data were not retained," and that the researcher was sanctioned by ORI even though he "denies all allegations of scientific misconduct and contends that some of his original data is missing."

Exercise 7.2
Read and discuss with a colleague the following Finding of Scientific Misconduct: https://www.ncbi.nlm.nih.gov/pmc/articles/PMC4259619/. What aspects of this report do you find surprising? What impact might the sanctions imposed have on a researcher's career? Do you agree with the practice of publicly disclosing these findings and naming the researcher involved? What impact might the fraud in this case have had on other researchers, doctors, or patients? Given the impact of the fraud, was the severity of the imposed sanctions appropriate?

7.2   Why Keep Records?

Accurate records are central to any investigation of scientific misconduct, yet such investigations are rare. Not surprisingly then, the ability to defend against an accusation of misconduct is not the primary reason researchers keep records, and this potential concern should not dominate our discussion of recordkeeping. A discussion of what records to keep and how best to do so begins with a consideration of what information will be needed in the future and why.

Exercise 7.3
First, list reasons why physicians use a medical chart. Compare your list with one or more colleagues and add to your list as needed until you believe it is complete. Second, make a similar list of reasons that researchers at a medical device company record information on design and development. Compare this list with your list for medical charts. How many of the reasons for keeping records appear on both lists? Third, list reasons why a researcher working in academia records research methods or findings. Are there any reasons unique to this third list?

7.2.1   Medical Records

Although your list may differ, commonly cited reasons for maintaining a detailed medical record are immediate transfer of information, long-term transfer of information, training medical students and residents, and legal documentation. Examples of immediate information transfer include a physician recording an order in a chart that another member of the hospital staff must execute later in the day, or a resident who is called in the middle of the night to examine a patient deciding an appropriate course of action based in part on a review of the patient's chart. Because many different people come in contact with each patient during a typical day in a hospital, a smooth transfer of information can literally be the difference between life and death.

Availability of an accurate longer-term medical history can be equally important to a patient's health. Diagnosing and treating a patient often depends critically on details of that person's medical history: previous illnesses and surgeries, current medical problems and medications, allergies, and so forth. Patients may not remember, or even know, all the details of the medical history, and it is unreasonable to expect physicians to remember the detailed histories of thousands of patients under their care. Consequently, a well-documented record of each patient's medical history is not only essential to the accurate exchange of information between physicians, it is also critical as an accurate, detailed supplement to each physician's memory.

Perhaps less obvious, good recordkeeping can be useful in training medical students, nurses, and other health care professionals. Most entries in medical records have very specific formats. Learning and using these specific formats is integral to learning the thought process associated with medicine. One usually records a detailed medical history and the results from a physical examination using a form that lists standard questions and aspects of the examination. Recording the same information for each patient helps students learn the essential components of a good examination; they soon begin to ask the questions and perform the examination in the same order each time, which helps ensure that they do not miss anything. Another common entry in hospital charts is the SOAP note, an acronym for "subjective, objective, assessment, and plan." Organizing daily updates under these four headings encourages a particular thought process: gather the information, think about what it means, then decide what to do.

Finally, it is no surprise that medical charts serve as an important legal record of what happened to a particular patient and why. In fact, most respondents to Exercise 7.3 place this first in their list. Unfortunately, this legal function can conflict with the training function discussed above, as when hospitals bar medical students from adding to a patient's chart for fear that an erroneous assessment or plan, even if corrected later by the supervising physician, could increase vulnerability in a lawsuit.

7.2.2 Industry Research Records

Your second list from Exercise 7.3 may not differ much from the first list. Laboratory notebooks, whether hardcopies or electronic, kept by employees of a medical device company serve many of the same functions as a medical chart. Multiple technicians might record results from a series of tests for review and compilation by their supervisor the next day (short-term transfer of information). Technicians may consult their records when performing the same tests a month later to make sure they set them up exactly the same way (long-term transfer of information). Asking new employees to follow a specific structure for recording data from a particular test can help them learn how to perform that test (training). Finally, approval of new drugs or devices by the U.S. Food and Drug Administration requires stringent recordkeeping (legal documentation); such records are essential for the survival of pharmaceutical and device companies.

If you or a colleague with whom you discussed Exercise 7.3 has worked in industry, the topic of cosigning likely surfaced in your list. Many industrial research facilities require a supervisor to review and cosign laboratory records at the end of each day. This requirement can assist many of the functions of recordkeeping discussed above. If a test result is surprising, a supervisor can learn about the result and take appropriate action immediately: check the equipment, schedule a repeat test, or discuss the findings with his or her boss (information transfer, quality control). Daily review also provides an excellent opportunity for feedback on how best to perform the test or record the results. Finally, for companies that depend critically on regulatory approval of their products, cosigning not only helps to ensure proper performance, it also ensures all tests and procedures are appropriately documented (legal).

7.2.3   Academic Research Records

By now, the pattern should be apparent; recordkeeping serves similar functions in diverse disciplines and settings. Academic researchers use records to transfer information amongst members of a group, as a long-term record of what was done and how it was done, to help train students to perform and record their work, and as a legal record. Because most academic researchers are more interested in publishing journal articles than protecting themselves against product liability suits, the need for long-term documentation tends to dominate in academic practice. Nevertheless, answers to Exercise 7.1 usually reveal that many research groups do not keep adequate records, even to meet the basic goal of documenting what research was performed, when, and by whom. One of many interesting findings that emerged from a 1993 survey of postdoctoral research fellows at the University of California San Francisco was that fellows with an M.D. degree were significantly more likely to keep laboratory records [at that time in ink in a permanently bound research notebook] than were those with a Ph.D. degree (Eastwood et al., 1996).

Exercise 7.4
Design a recordkeeping policy for your research group. What should be recorded and where? Should cosignatures be required? If yes, who should cosign and how

often? Should cosigning or other rules of recordkeeping differ for different members of the group (e.g., undergraduates, graduate students, postdoctoral fellows)? How should new members who join the group be instructed in keeping records? Who should be responsible for ensuring that rules are followed? What should be the consequences for a group member who fails to keep appropriate records? What should happen to a member's records when they graduate or leave the group? Finally, are any special rules needed for electronic data? How does your policy compare to your own current practices in your research?

7.3   Recordkeeping in a Digital Era

Medical charts and laboratory notebooks originated as physical, paper records. While bulky and inconvenient in many ways, paper records preserve a sequential history of events that can be valuable. Consider a spreadsheet or data file containing results from a dozen experiments performed over a 6-month period. If you had produced this data file and were asked to verify that you actually performed the experiments, what proof could you offer? The data file itself is of little use; one can artificially generate data and save the results in the desired format. The operating system will likely display the date you created the file and the date you last modified it, but these dates can be manipulated by changing the computer's clock. If you had kept every version of the file as you entered data from each new experiment, paging through these versions would be more convincing, but this trail of different versions of the file could also be fabricated. Having different versions of the file plus a physical notebook showing data acquisition on dates that match those for the files would be better, especially if the notebook had been cosigned by a supervisor at intervals over the period in question.

That most research today relies heavily on digital records presents enormous challenges for ensuring integrity of the science and engineering. While researchers rarely need to prove that they performed specific experiments, they often need to revert to an older version of software or refer to computer files that have been deleted or altered. Consider this fairly common dilemma from the perspective of the director of a research laboratory at a university. A new student begins working for you, taking over a project from a previous student who graduated. Her project involves imaging cells and analyzing the images with custom-written software that is still under development in your group, and her early results appear to contradict findings by her predecessor. To diagnose whether the differences arise from differences in isolating the cells, imaging them, using the software, or even changes in the software itself, you will need a trail of previous images, versions of the software, and well-annotated analyses. Few laboratories proactively anticipate such situations in their recordkeeping or procedures for data backup, yet most will experience them.

Given the availability of open-source tools and inexpensive digital or cloud storage, any research laboratory can construct a good system of digital storage that addresses the challenges described above. These systems will differ, but key principles include: (1) storing locked original images or data files generated by experiments; (2) making periodic backups of laboratory computers to the cloud or a backup system located outside the laboratory; (3) periodically saving snapshots of files that are evolving (such as working notes, summary spreadsheets, or software) with a filename that indicates the date that version was current; and (4) implementing a revision control system for software and/or other evolving files. If you are not familiar with revision control systems, we encourage you to read about them. Developed by the software industry to allow multiple people to work simultaneously on different aspects of a program, integrating changes made by different group members in an orderly way while tracking previous versions so that problematic changes can be undone, these systems address many of the same challenges faced by a typical research laboratory. Common platforms such as Google Docs now make some of these features available to teams working on a range of documents, not just software.

Exercise 7.5

Return to the recordkeeping policy you designed for your research group in Exercise 7.4. Provide more specific instructions for the primary types of digital files and records used in your group. What would your group need to change to implement your plan? If you would need additional storage, estimate the amount of space you would need and determine how much this would cost. If you believe that your group should implement software revision control, identify open-source and commercial products that might meet your needs. If you feel strongly that changes are needed, consider discussing potential changes and options with your colleagues.

7.4   Ownership of Data and Other Research Products

One of the many interesting questions in science is who owns the results. Modern trends, such as the drive to commercialize products of university research and to increase public access to data from federally funded research, highlight this question of ownership. Debates about ownership of and access to the results of scientific and engineering research have important consequences for individuals, universities, companies, publishers, the government, and the public at large.

Earlier in this chapter, we compared medical charts to laboratory notebooks when considering why we should keep records. This comparison also provides an interesting entry into the question of ownership. Who owns your medical record and what conceptual tests for ownership does your answer suggest? The information is about you and you can request a copy of the records, which suggests some level of ownership. Nevertheless, you cannot alter or destroy the original medical record, which suggests that someone else shares ownership. The physicians, nurses, and other medical personnel who produced the record can write in it and read it, but they cannot obtain a personal copy to take home. What about the insurance company who paid for your care? Does paying for the associated diagnostic tests and office visits give the company any stake in ownership of the information?

Exercise 7.6

Based on your experience in research and the analogy to a medical chart outlined above, discuss why each of the following parties should or should not be considered owners of scientific data (not the resulting publications or patents) that result from a research project funded by the federal government using tax dollars and performed at a private university:

1. The principal investigator of the study
2. The students, technicians, and fellows who performed the experiments
3. The university
4. The federal government
5. The public/taxpayers

Now, for each of the people, groups, or institutions listed above, specify appropriate levels of access to the original data. In other words, who should be able to acquire copies of all the data, to alter those data, or to use them to write papers or submit patents? Are there people who should not be considered co-owners but who should still be granted some access? Explain.

Before 1980, the federal government owned patent rights to any discovery made with federal funding. The simple rationale for this policy was that inventions generated with public funds should belong to the public. This same rationale underlies the NIH requirements for sharing of data and other research products discussed in Chap. 4. As suggested by the exercise above, however, the general question of ownership is not simple. As with data and publications, generating patents not only requires funding but also knowledge, ingenuity, hard work, equipment, space, and other resources. Hence, faculty, students, the university, and the government might all credibly claim at least partial ownership of a patent based on publicly funded research performed in a university laboratory. In 1980, the U.S. Congress passed the Bayh–Dole Act, which granted universities and small businesses the rights to patents arising from their federally funded research, on the argument that transferring ownership to companies and universities would promote greater dissemination and utilization of innovations generated from federal funding. From an ownership perspective, this policy was also more balanced than the one it replaced—it granted patent rights to universities, required universities to share royalties with the inventors, and retained limited rights for the government.

This shift in policy dramatically changed the technology transfer landscape. Most research universities operate substantial technology transfer offices, securing and licensing patents covering a range of ideas and products derived largely from federally funded research. A network of consulting, research, and technology transfer agreements links universities to companies, and ownership of intellectual property is frequently the critical concern during the negotiation of such agreements. Professors routinely form "spin-off" or "start-up" companies to translate their discoveries into products. In a complete reversal from the era before Bayh-Dole, the most critical current questions regarding publicly funded technology involve how to protect the public's interest rather than how to promote commercialization.

7.5 Fabrication, Falsification, and Manipulation

A review of current cases on the ORI website shows that some scientists are willing to fabricate or falsify data to obtain grants and publish papers. Surveys and other studies suggest that these high-profile fraud cases are the tip of an iceberg. While outright fabrication of data and images is relatively rare, many other types of manipulation are surprisingly common. Often termed "questionable research practices," these less blatant but much more common actions by individual researchers appear to pose a significant risk to the integrity of the scientific enterprise at a time of diminished public confidence in science and scientists. While we consider it unlikely that reading a book chapter will fundamentally change someone's willingness to act unethically, considering in advance some decisions and consequences you may face and deciding in advance how you plan to act may help you remain consistent to your own values as you build your career.

7.5.1 Retaining or Discarding Data

Consider a student who has carefully collected and plotted a dozen data points on an x–y graph, then fitted a line through the data. Fabricating extra data that lie near the best-fit line would improve the r^2 value typically reported to indicate how well the line fits the data; classifying a few points far from the best-fit line as outliers and discarding them would similarly improve the reported fit. Why is it that every scientist or engineer would consider fabricating extra data points to be fraudulent, while many would at least consider discarding outliers?

The *Nature* article titled "Scientists Behaving Badly" reported findings from a survey of more than 3000 NIH-funded scientists regarding the self-reported frequency of behaviors identified as concerning by focus groups of researchers (Martinson et al., 2005). While only 0.3% admitted to falsifying research data within the previous 3 years, 15.3% admitted to "dropping observations or data points from analyses based on a gut feeling that they were inaccurate." The only more frequent offense (27.5%) was "inadequate record keeping related to research projects."

A series of related studies over more than 35 years suggest that inappropriate handling of data is widespread and has not changed much despite increased attention to training in Responsible Conduct of Research (RCR). Fanelli (2010) conducted a meta-analysis of studies published from 1987 to 2008 and computed a pooled weighted estimate of 2% for self-reported fabrication and falsification of data and 14% for the percentage of scientists who said they had observed fabrication or falsification by a colleague. More recent surveys of scientists and economists continue to find ranges of self-reported (0–4%) and observed (4–24%) falsification consistent with Fanelli's review (Godecharle et al., 2018; Gopalakrishna et al., 2022; Necker, 2014; Tijdink et al., 2014).

Unfortunately, self-reported and observed levels of data manipulation are much higher than those of outright fabrication. Surveys that repeat the question from Martinson et al. (2005) about dropping observations based on a gut feeling or that ask related questions about manipulating data to confirm a hypothesis find that up to 21% of scientists admit to these behaviors (Parlangeli et al., 2020) and 40% or more say they have observed them (Godecharle et al., 2018; Tijdink et al., 2014). These numbers are also consistent with studies of trainees that indicate 15% or more are willing to select or omit data to increase the chances of publishing a paper or receiving a grant (Eastwood et al., 1996; Kalichman et al., 1992).

Exercise 7.7

Formulate criteria for appropriately discarding a data point, observation, or study. Compare your draft policy with that of one or more colleagues and revise until you believe that your policy could be implemented in your group. Next, find a data set collected in your group that you know contains noise or errors and test your policy by applying it to the data set. How many points would you discard, if any, and what justification would you give? As a final step, discuss your analysis with colleagues, then with a senior colleague or mentor. Submit a three-page summary of your conclusions.

An Internet search will return many different criteria and algorithms for identifying "outliers" in a data set; consider, for example, "Best-practice recommendations for defining, identifying, and handling outliers" by Aguinis et al. (2013). Exclusions must combine good experimental and statistical practices with an understanding of the overall study and potential impact of those exclusions. In general, exclusion criteria should be established before performing the study and should depend on something other than the values obtained. Excluding data collected on a single day that appear to be outliers, based on the rationalization that the equipment must have been miscalibrated, is a dangerous practice. By contrast, performing a calibration at the end of each day and excluding from analysis all measurements taken on days where the calibration fell outside a preset tolerance is good technique. Discussing criteria for excluding data before performing a study can help to ensure that you have enough information to make good decisions later. Other good practices include discussing exclusions with your mentor or colleagues before making them, disclosing exclusions and the associated justifications in publications, and remembering that if something does not "feel" right, it probably is not. Finally, we note that paying close to attention to outliers—rather than simply ignoring them—can have important scientific benefits. If they did not arise from technical errors, understanding why the system you are studying sometimes behaves so differently from your expectations could provide a clue to deeper understanding of that system.

7.5.2   Image Manipulation

The importance of digital images in many types of research raises additional questions about the degree of processing or manipulation that is appropriate in a given situation. Consider a gel such as a Western blot, where a series of stained bands indicate relative amounts of certain proteins (rows) in different samples (columns or lanes). Is there anything unethical about cutting and pasting from several images of gels run on different days to create a composite image showing results from certain samples side by side? What if one or more of those gels was underexposed? The resulting image would then be lighter than expected and this problem could be corrected in one of two ways: reimage the gel with a longer exposure time or adjust the image contrast and brightness using an image processing program. Are these two options really the same, or is manipulating the image fraudulent?

The rise of Adobe Photoshop and other similar programs made it much easier to edit images in ways that go far beyond adjusting contrast or brightness, such as removing or adding bands in particular lanes to change the apparent quality of supposedly 'raw' data. In 2004, the editors of the *Journal of Cell Biology* called attention to the risks of image manipulation in a paper titled, "What's in a picture: the temptation of image manipulation" (Rossner & Yamada, 2004), which included a clear policy for all images in their journal:

> No specific feature within an image may be enhanced, obscured, moved, removed, or introduced. The grouping of images from different parts of the same gel, or from different gels, fields, or exposures must be made explicit by the arrangement of the figure (e.g., using dividing lines) and in the text of the figure legend. Adjustments of brightness, contrast, or color balance are acceptable if they are applied to the whole image and as long as they do not

obscure or eliminate any information present in the original. Nonlinear adjustments (e.g., changes to gamma settings) must be disclosed in the figure legend.

Rossner and Yamada (2004)

Rossner and Yamada went on to list various types of editing that could be performed on images, explaining which were acceptable and unacceptable according to the journal policy and providing examples of various image manipulations inspired by cases they had observed.

Elizabeth Bik, a microbiologist working at Stanford University, became interested in scientific fraud and developed a hobby of finding images in research papers that appeared to be manipulated (Shen, 2020). In 2016, she published the results of her own visual screening of images in more than 20,000 papers published between 1995 and 2014 that contained at least one Western blot; two co-authors confirmed her analysis (Bik, 2016). Bik focused on image duplication, a specific type of image manipulation where a region of an image is copied and (1) re-used in another figure to represent a supposedly different experiment, (2) pasted into another part of a figure, sometimes after rotation or reflection, or (3) altered and pasted into another part of a figure. Reportedly, 3.8% of the papers screened contained problematic figures, with a lower prevalence from 1995 to 2004 and rates $\geq 4\%$ from 2005 to 2014. Interestingly, the *Journal of Cell Biology*—with its long-standing policy against image manipulation—contained the lowest percentage of papers with image manipulation among the 40 journals included in the screen. Bik and colleagues also showed that once they identified a paper with manipulated images, additional publications by the same authors were much more likely to contain similar problems, suggesting that authors willing to alter images are willing to do so repeatedly.

A number of software tools are available that can help detect image manipulation. One useful resource is the Forensic Tools website from the U.S. Department of Health and Human Services Office of Research Integrity (https://ori.hhs.gov/forensic-tools). This website provides downloadable applications and Adobe Photoshop add-ins, as well as sample images useful for learning how to apply the forensic tools. While available to anyone, these tools are particularly useful for university committees investigating allegations of fraud, and for journal reviewers who suspect problems with an image in a paper they are evaluating.

7.5.3   Data Obtained from Others

Findings of misconduct from the Office of Research Integrity frequently feature cases of undergraduates, graduate students, postdoctoral fellows, and principal investigators fabricating data or manipulating images, and even laboratory members interfering with experiments performed by colleagues, in order to prevent detection of prior fraud. If you are currently a student or postdoctoral fellow, you may be focused primarily on learning how to perform experiments and analyses correctly and to document that work. As you move forward in your career, you will begin to get more of your data secondhand from others, including students, employees, or collaborators. It is worth considering what steps you should take so that you can be confident in these data when you include them in your own papers and grants. Providing good training and enforcing good recordkeeping within your own group can certainly help, as can reviewing not just summarized or processed data but also the original files or records. Having a second person replicate selected experiments can improve confidence in specific results while also guarding against errors and problems with reproducibility (see below), as well as ensuring continuity of methods as members join or leave the group.

7.6   Plagiarism and Self-Plagiarism

The recent history of plagiarism in scientific literature follows an arc similar to that of image manipulation. Digital access to an increasing volume of scientific literature initially enabled plagiarism, yet also facilitated the development of increasingly sophisticated tools to detect it. Errami and Garner (2008) used the text-matching software eTBLAST to search Medline abstracts and found that the fraction of suspected duplicates increased steadily from 1975 to 2005 to a little over 1%. Around the same time, full-text searches of papers in arXiv.org using a custom algorithm identified approximately 10% of the papers examined as duplicates (Sorokina, 2006). Today, full-text searches against very large document databases are possible using commercial plagiarism-detection software such as Turnitin or iThenticate. As one example, Horbach and Halffman (2019) used Turnitin to look for matches to more than 900 papers by Dutch researchers published after 2010; even after excluding duplicate passages that were referenced appropriately, more than 5% of papers contained text that was at least 10% identical with a previous publication.

Such studies consistently report extremely low rates of true plagiarism, defined as duplication of substantial blocks of text by a different set of authors. Instead, the vast majority of overlapping text identified in the literature is categorized as "self-plagiarism" or "text recycling," in which authors utilize identical sentences or longer blocks of text from their own previous publications. There has been considerable debate in the scientific community about the extent to which such re-use should be allowed. Necker et al. (2014) found that nearly 25% of scientists they surveyed admitted "copying from your own previous work without citing," and that their respondents considered this to be at best a minor infraction (rated 2.5 on a scale where 1 = not at all justifiable and 6 = highly justifiable). Other surveys found that as many as 5% of researchers admit to publishing the same data in two or more publications, and between 10 and 40% report observing colleagues doing the same (Godecharle et al., 2014; Martinson et al., 2005; Tijdink et al., 2014). Some of the variation in reported prevalence likely relates to the exact phrasing of the question. For example, Parlangeli et al. (2020) reported that nearly half of Italian researchers they surveyed agreed that they had "published the same research in different media (proceedings, journal articles, book chapters)," but this phrasing could plausibly include appropriate development of an early abstract into a full-length journal publication or summarizing previously published work appropriately in a review article or book chapter.

The Committee on Publishing Ethics (COPE) develops guidelines, policy statements, and flowcharts designed to help publishers, editors, and reviewers achieve a high standard of ethics in publishing (https://publicationethics.org/). We recommend their guidelines on text recycling to readers who have questions about how to handle these issues as authors or reviewers. Some of the key points include: (1) limited repetition of text from your own prior publications is likely acceptable in the Methods section, with appropriate citation; (2) by contrast, text repetition in the Results and Conclusions sections is rarely justified since an original paper should test new ideas and contribute new findings; (3) limited re-use of some results from a pilot study or abstract may be appropriate, as long as this is clearly disclosed in the article and cover letter. Two specific types of "self-plagiarism" that are never acceptable include using the same figure in a second paper without permission from the original journal (which violates copyright) as well as submitting the same manuscript to two journals simultaneously (which violates the journal submission agreement).

With this in mind, we encourage the reader to consider carefully the potential reuse of text from and in the Methods section and to discuss this with coauthors. In an era wherein rigor and reproducibility are of paramount importance, it should be expected that certain methods remain the same across multiple studies; indeed, such practice enables direct cross-comparisons of results, which promises to increase understanding. If the methods remain the same, should not the description of these methods also remain the same? One way to address this expectation is to note directly that the methods have been established previously (with proper citation) and to summarize the methods briefly. If the journal requests more detail on the methods, then reuse of appropriate text can be noted directly via quotation marks. A similar situation can arise with the potential reuse of previously reported data. Again in the spirit of increased rigor and reproducibility, it can be useful to compare directly the current data with previously reported (published) data in a figure or table. This can be accomplished simply by using different symbols, fonts, or notations accompanied by proper citation. Most importantly, remember that full disclosure (transparency) is always the best approach.

By contrast, a situation in which it is tempting to recycle text inappropriately is when a student new to a topic is writing his or her first paper. The starting point for the study may well be prior work from the laboratory, and the Introduction section of previous papers from the group might contain many of the same points needed in the new Introduction. One tip to help make the paper your own is to construct a bulleted outline of the points you want to make in each section and to discuss the outline with your mentor and co-authors. There are many advantages to this approach, including that it is easier to see the overall flow of ideas in a shortened format, and therefore easier to re-arrange, add, and delete ideas until everyone is satisfied. Another advantage is that abstracting concepts from prior papers into bullets and then expanding them into full sentences and paragraphs in your own words ensures that the text is original.

7.7 Retractions

One of the remarkable features about the Slutsky case was that some journals were unwilling to retract papers identified as fraudulent unless the authors (including Dr. Slutsky) agreed. This aspect of the publishing landscape has changed considerably. Grieneisen and Zhang (2012) found documented retraction notices for less than 50 papers per year from 1980 to 2000 but more than 500 papers per year by 2010. Some of this increase was likely due to a general increase in the volume of scientific publications (Steen, 2013), yet another reason may include the increasing pressure to publish in so-called high impact journals (Fang 2011). Additional evidence supports the idea that journals and others in the scientific community were becoming better at detecting and correcting problems. Publishers retracted articles faster in the decade 2002–2012 relative to earlier periods (Steen, 2013), and the number of journals that published retractions increased tenfold from 1997 to 2016 (Brainard & You, 2018).

In 2010, health journalists Adam Marcus and Ivan Oransky launched the blog Retraction Watch (https://retractionwatch.com/) to track trends in how journals handle fraud, error, and other problems; in 2018, they unveiled a searchable database of retractions. An analysis of their database showed that retraction rates rose through 2010, as reported by others, but then appeared to level off at about 4 papers for every 10,000 published (Brainard & You, 2018). Plagiarism accounted for a much larger fraction (more than 30%) of all retractions in recent years, perhaps reflecting the widespread availability of software to detect it. Overall, the Retraction Watch data and earlier studies agree that scientific fraud accounts for less than half of all retractions, and that a surprisingly large share of fraud retractions stems from major cases involving a handful of investigators (Brainard & You, 2018, Grieneisen & Zhang, 2012, Steen 2013).

7.8 Reproducibility

Many scientists are attracted to our profession in part by the belief that our job is to search for truth. Early in our training, many of us learned an idealized version of the scientific process wherein one laboratory makes a new discovery and publishes it, others attempt to replicate that finding, and the new finding is accepted as 'truth' only after the scientific community has reached consensus. This process should ensure that new discoveries are durable and reproducible—any lab that performs the same experiment under the same conditions will get the same result. Future work builds on the foundation of previously confirmed knowledge, and as a community we gradually build our understanding of the world, moving ever forward.

Any professional engaged in the practice of science quickly learns that reality is messier than the idealized version outlined above. Experiments are often complex, and even when authors make their best effort to provide appropriate methodological details in a published paper, reproducing published studies can be difficult. The same experimental finding can also be interpreted in different ways, so agreeing on the result of an experiment does not necessarily lead to agreement on its meaning. As in any field, people make mistakes. And as the body of scientific knowledge grows, a number of real-world constraints make it ever less likely that the community will attempt to replicate individual published results.

One of the most fascinating emerging developments in the study of science itself is the realization that our current systems for funding and publishing science virtually guarantee that findings from a substantial portion of published studies cannot be reliably reproduced, even when everyone involved acts ethically and according to accepted scientific standards.

One aspect of this dilemma arises because most scientists (and similarly journals) prefer to focus on positive results ('new' findings). Imagine that 20 laboratories are simultaneously testing the ability of the same new drug to lower blood pressure in rats. By convention, scientists generally set a threshold of P value $= 0.05$ for accepting a statistical test as significant, meaning that the chance of a false positive result is only 5%. If the drug has no real impact on blood pressure, we might expect 19 groups to measure no significant effect and one group to find a (false positive) significant effect by random chance. The group with the positive result would likely write a paper describing their new finding while groups with the negative findings might choose to move on to the next study rather than trying to publish a negative result. In that case, the only paper published in the scientific literature will be the one with the misleading result! There are many variations on this theme. For example, because papers with negative results are viewed as less exciting, one of the negative studies might be published in a lower-profile journal than the study with the positive result. In that case, both studies can be discovered through a literature search, but the positive study might be given more credence.

Another aspect of this dilemma arises from the understandable focus on innovation by funding agencies (recall Chap. 4). Although we all expect science to be reproducible, in practice funding agencies are reluctant to fund research that appears to duplicate previous work. They are also less likely to fund similar work by multiple investigators. As a result, our hypothetical scenario of 20 labs testing the same drug in parallel is unlikely in the real world. Rather, it is possible that two or three laboratories might be testing the drug, and that a false positive result could emerge from one of those studies. At that point, it becomes difficult for other laboratories to obtain grant funding to study what is viewed as a settled question, and the drug might move further down the development pathway even though a larger number of studies and full disclosure of their results would have shown the drug to be ineffective.

In 2011, a group of psychology researchers set out to test the reproducibility of experimental studies published in three specific psychology journals in the year 2008 (Open Science Collaboration, 2012). They established a formal protocol for choosing and replicating the studies that mandated designs with high statistical power, coordination with the original authors to ensure that the experimental protocols matched the original studies as closely as possible, and other measures intended to ensure that replication attempts were performed and reported in a rigorous, open, and transparent manner. In 2015, this group published the results of 100 replication attempts. While 97 of the original studies reported a P value < 0.05 for the specific experiment being replicated, only 35 of the replication studies found a P value < 0.05 (Open Science Collaboration, 2015). When the strength of the reported relationships was expressed as a correlation coefficient, the average strength of the relationship ('effect size') dropped by half in the replication studies.

The authors of this landmark study were careful to point out that there is no way to know which of the original (or even replication) results are 'false positives' or 'false negatives,' or to reach a definitive conclusion about why the replication results were so different from the original studies. Yet overall, there was a clear trend that careful attempts to replicate published studies consistently find weaker evidence for published results than the original papers, and the authors noted that systematic bias towards selecting, reporting, and publishing positive results would be a plausible explanation for their findings (Open Science Collaboration, 2015).

As the Open Science Collective was examining replication experimentally, the physician and epidemiologist John Ioannidis and his colleagues were raising alarms about bias and the focus on positive results from a statistical perspective. In his 2005 essay, "Why Most Published Research Findings Are False," Dr. Ioannidis considered how statistical power, bias, and the search for low-probability events in analyses such as Genome-Wide Association Studies (GWAS) intersect with the standard threshold P value < 0.05 for statistical significance and the tendency to report and publish positive results (Ioannidis, 2005). Among other situations, Dr. Ioannidis noted that in a study where a single hypothesis is tested and is equally likely to be true or false, the significance level alpha $= 0.05$, and the power (1-beta) $= 0.80$, the expected frequency of true and false positives and true and false negatives is as shown in this table familiar to most who have taken a statistics course:

TRUE POSITIVE	FALSE POSITIVE
FALSE NEGATIVE	TRUE NEGATIVE

0.400	0.025
0.100	0.475

In this case, there is a 50% chance that the hypothesis is true and an 80% chance of detecting a true relationship, thus the overall probability of a true positive result is 0.40. In parallel, there is a 50% chance that the hypothesis is false and a 5% chance of finding P < 0.05 despite the hypothesis being false, so the overall probability of a false positive is 0.025. By contrast, when screening 100 genes to find a single gene that is truly related to a disease of interest, the table looks like this:

TRUE POSITIVE	FALSE POSITIVE		0.0080	0.0495
FALSE NEGATIVE	TRUE NEGATIVE		0.0020	0.9405

In other words, when most hypotheses being tested are likely to be false, false positives are much more likely that true positives. Fields such as genomics that deal with large screens have developed statistical methods to account for the large number of comparisons performed in such tests. Yet this example serves as one more caution about the unintended consequences of focusing on and publishing only positive results.

7.9   Responsibility

As discussed in multiple Chapters, modern science and engineering is increasingly complex and multidisciplinary. Seldom does a single investigator collect, analyze, or scrutinize all of the findings, whether experimental or computational. There is, therefore, a critical need for all members of the research team to take responsibility for their part of the study and to document and communicate the safeguards taken to ensure that all of their reported findings are properly recorded, reliable, and reproducible.

References

Aguinis, H., Gottfredson, R. K., & Joo, H. (2013). Best-practice recommendations for defining, identifying, and handling outliers. *Organizational Research Methods, 16*(2), 270–301. https://doi.org/10.1177/1094428112470

Bik, E. M., Casadevall, A., & Fang, F. C. (2016). The prevalence of inappropriate image duplication in biomedical research publications. *mBio, 7*(3), e00809-16. https://doi.org/10.1128/mBio.00809-16.

Brainard, J., & You, J. (2018, October 25). What a massive database of retracted papers reveals about science publishing's 'death penalty.' *Science News.* https://doi.org/10.1126/science.aav8384.

Eastwood, S., Derish, P., Leash, E., & Ordway, S. (1996). Ethical issues in biomedical research: Perceptions and practices of postdoctoral research fellows responding to a survey. *Science and Engineering Ethics, 2*, 89–114. https://doi.org/10.1007/BF02639320

Engler, R. L., Covell, J. W., Friedman, P. J., Kitcher, P. S., & Peters, R. M. (1987). Misrepresentation and responsibility in medical research. *New England Journal of Medicine, 317*, 1383–1389.

Errami, M., & Garner, H. (2008). A tale of two citations. *Nature, 451*(7177), 397–399. https://doi.org/10.1038/451397a

Fanelli, D. (2010). Do pressures to publish increase scientists' bias? An empirical support from US States data. *PLoS ONE, 5*(4), e10271. https://doi.org/10.1371/journal.pone.0010271

Godecharle, S., Fieuws, S., Nemery, B., Dierickx, K. (2018). Scientists still behaving badly? A survey within industry and universities. *Science and Engineering Ethics, 24*, 1697–1717. https://doi.org/10.1007/s11948-017-9957-4

Gopalakrishna, G., Ter Riet, G., Vink, G., Stoop, I., Wicherts, J. M., & Bouter, L. M. (2022). Prevalence of questionable research practices, research misconduct and their potential explanatory factors: A survey among academic researchers in The Netherlands. *PLoS ONE, 17*(2), e0263023. https://doi.org/10.1371/journal.pone.0263023

Grieneisen, M. L., & Zhang, M. (2012). A comprehensive survey of retracted articles from the scholarly literature. *PLoS ONE, 7*(10), e44118. https://doi.org/10.1371/journal.pone.0044118

Horbach, S. P. J. M., & Halffman, W. (2019). The extent and causes of academic text recycling or 'self-plagiarism.' *Research Policy, 48*(2), 492–502. https://doi.org/10.1016/j.respol.2017.09.004

Ioannidis, J. P. A. (2005). Why most published research findings are false. *PLoS Medicine, 2*(8), e124. https://doi.org/10.1371/journal.pmed.0020124.

Kalichman, W. M., & Friedman, P. J. (1992). A pilot study of biomedical trainees' perceptions concerning research ethics. *Academic Medicine, 67*(11), 769–775. https://doi.org/10.1097/00001888-199211000-00015

Martinson, B. C., Anderson, M. S., & de Vries, R. (2005). Scientists behaving badly. *Nature, 435*, 737–738. https://doi.org/10.1038/435737a

Necker, S. (2014). Scientific misbehavior in economics. *Research Policy, 43*(10), 1747–1759. https://doi.org/10.1016/j.respol.2014.05.002

Open Science Collaboration. (2012). An open, large-scale, collaborative effort to estimate the reproducibility of psychological science. *Perspectives on Psychological Science, 7*, 657–660. https://doi.org/10.1177/1745691612462588

Open Science Collaboration. (2015). Estimating the reproducibility of psychological science. *Science, 349*(6251), aac4716. https://doi.org/10.1126/science.aac4716.

Parlangeli, O., Guidi, S., Marchigiani, E., Bracci, M., & Liston, P. M. (2020). Perceptions of work-related stress and ethical misconduct amongst non-tenured researchers in Italy. *Science and Engineering Ethics, 26*(1), 159–181. https://doi.org/10.1007/s11948-019-00091-6

Rossner, M., & Yamada, K. M. (2004). What's in a picture? The temptation of image manipulation. *Journal of Cell Biology, 166*(1), 11–15. https://doi.org/10.1083/jcb.200406019

Shen, H. (2020). Meet this super-spotter of duplicated images in science papers. *Nature, 581*, 132–136. https://doi.org/10.1038/d41586-020-01363-z

Sorokina, D., Gerke, J., Warner, S., Ginsparg, P. (2006). Plagiarism detection in arXiv. In *Sixth International Conference on Data Mining (ICDM'06)*. https://doi.org/10.48550/arXiv.cs/0702012.

Steen, R. G., Casadevall, A., & Fang, F. C. (2013). Why has the number of scientific retractions increased? *PLoS ONE, 8*(7), e68397. https://doi.org/10.1371/journal.pone.0068397

Tijdink, J. K., Verbeke R., Smulders Y. M. (2014). Publication pressure and scientific misconduct in medical scientists. *Journal of Empirical Research on Human Research Ethics, 9*(5): 64–71. https://doi.org/10.1177/1556264614552421

Review and Free Exchange of Ideas

8

8.1 Context

Throughout the first seven chapters we have encouraged the reader to develop a thoughtful, individual style of communication that respects common conventions and adheres to ethical standards. Such communication is fundamental to the advancement of science and engineering. Toward this end we have also emphasized the importance of both appropriate attribution, most often via proper citation of documented works by others, and meticulous record keeping, which promotes the integrity of the work. In this way, we have encouraged you, the reader, to become a responsible contributor to the advancement of your field.

For advances in science and engineering to improve the human condition, fundamental ideas, discoveries, and inventions must be translated into everyday use. Recall from Chap. 7 that there evolved a legal system to protect intellectual property and encourage translation. Fundamental to this system is the *patent*, which derives from the Latin meaning "lying open." That is, patent protection documents rightful ownership and thereby allows an individual to openly disclose and develop an idea, discovery, or invention.

> **Exercise 8.1**
> Noting that universities are charged with the mission of discovery and dissemination of knowledge, it may not be surprising that not that long ago, in 1970, only three major research universities devoted at least one half-time staff position to technology transfer and research universities as a group secured only ~150 patents (Sampat, 2006). Most universities consciously confined their activities to the generation and free exchange of knowledge; they avoided the business aspects of translating that

© The Author(s), under exclusive license to Springer Nature Switzerland AG 2024   141
J. D. Humphrey and J. W. Holmes, *Style and Ethics of Communication in Science and Engineering*, Synthesis Lectures on Engineering, Science, and Technology,
https://doi.org/10.1007/978-3-031-39125-5_8

knowledge into profitable products for fear that it would taint their academic missions. It is reported that, among others, Columbia University, Harvard University, Johns Hopkins University, The University of Chicago, and Yale University specifically prohibited the patenting of results from biomedical research. At that time, private companies performed most federally funded research and development, and patents derived from that research belonged to the federal government. University owned patents typically resulted from industry funded research and were designed primarily to protect against misuse of the technology. Licensing was assigned via independent foundations or corporations. Research the development of patents and technology transfer in American universities over the past five decades and describe associated advantages or disadvantages in a 5-page report.

In this Chapter, however, we conclude by addressing another aspect of the respect of ownership, and thus free exchange, of ideas that relates directly to three of the most common forms of communication in science and engineering—the submission of papers for possible publication, submission of proposals for possible funding, and presentation of ideas, findings, and open problems at a professional conference or invited seminar. In each case, one typically discloses, describes, and discusses in detail new ideas that are not patent protected—how then do we as a community encourage an ethical respect of the ownership of ideas that are presented for review or consideration prior to publication?

Fundamental to these activities is the role of individuals, often colleagues, in evaluating the work—anonymous reviewers, study sections or review panels, and judges of paper competitions among them. As with other aspects of science, peer review can raise diverse ethical questions. Some of these questions relate to our discussion in Chap. 7 about ownership of ideas, while others relate to reviewers who are in a position of power relative to the author of a manuscript or a grant application. Similar questions can arise in a less formal setting, when unpublished ideas are presented at a conference or seminar. Albeit not an exhaustive discussion, please consider the following while mentally placing yourself in the position of both author/presenter and reviewer/listener.

8.2   Peer Review

Science and engineering rely heavily on peer review, the evaluation of your work by others who have similar interests and expertise. Peer review is central to deciding which papers a journal will publish, which proposals agencies will fund, which patents will be issued, and which drugs and devices will be approved. Any system of peer review must balance the fact that colleagues who work in your field are best qualified to review your work against the possibility that those colleagues include former students, mentors, or collaborators as well as current collaborators and competitors. Evaluating and avoiding conflict of interest, either real or perceived, is integral to effective peer review.

Exercise 8.2

Imagine that you have been named editor of a new journal in your field. Formulate a policy stating how your journal will handle potential conflicts of interest when assigning reviewers. What constitutes a conflict of interest? Are all conflicts equal? How will conflicts be handled? Must any reviewer with a potential conflict decline to review, or are there some situations where declaring the conflict will suffice? Will you hire staff to search for potential conflicts or will you rely on reviewers to disclose conflicts? Will you allow authors to name reviewers they consider to be in conflict? What other measures will you take? Summarize your thoughts in a two-page paper.

8.2.1 Submitted Journal Articles

A journal editor who receives a manuscript for review usually begins by reading the abstract and scanning the paper to verify that it is appropriate for the journal. The next step is to assign reviewers. Editors typically identify reviewers from a variety of sources, including personal knowledge of investigators in the field, databases maintained by the journal, literature searches using keywords or title words from the manuscript, and the references cited within the manuscript. Some journals allow authors to suggest reviewers at the time of submission.

Selection of appropriate reviewers helps to ensure a fair and thorough review. Typically, editors try to assign several reviewers who are experts in the field but not associated closely with the authors or one another. Achieving this goal is not as easy as it sounds; it assumes that the editor knows professional relationships among those working in the field—who trained in which laboratories, who has collaborated with whom, who has published with whom. Most of this information can be found with enough research, but such research would be too time-consuming for every submitted manuscript. Hence, journals rely on multiple safeguards against conflict of interest. Securing multiple reviewers provides one safeguard, for it is less likely that multiple reviewers will have a conflict of interest or the same personal bias for or against a particular author. Some journals divide the task of assigning reviewers among associate editors or members of an editorial board to help ensure that assignments are based on detailed scientific and professional knowledge about subfields within a broader discipline. Many journals ask authors to list those whom they feel have a conflict of interest and will often exclude those reviewers.

Allowing authors to name potential reviewers for exclusion raises many interesting questions. As an author, you may have genuine concerns whether a particular competitor will judge your work more harshly than normal. Yet, are you willing to assert to the editor, often a senior colleague in your field, that this rival is incapable of judging your work fairly? Does having the potential to exclude particular reviewers tempt authors to try to avoid valid criticism that could help to improve the research and the paper? Ultimately, much relies on the judgment of the editor. Ideally, an editor who receives two very positive reviews and one negative, but unfair, review will read the reviews, recognize the lack of substance in the negative comments, and discount the unfair review when making a decision. If a journal receives a large number of submissions and is highly selective, however, an editor may simply average scores from all reviewers without resolving potential discrepancies; in such cases, one negative review may be enough to prevent acceptance. Fortunately, there are many journals in each field, hence there are always other opportunities to publish a high-quality paper. Recall from Chap. 3, however, that one should always take advantage of any opportunity to improve a paper via revision.

Assigning reviewers and integrating their feedback can be demanding, but it is usually straightforward from an ethical point of view. By contrast, reviewing an unpublished manuscript frequently raises difficult ethical questions. First, you must decide whether to accept the request to review. Generally, you should not agree to review the work of colleagues from your institution, of former students or mentors, or of current or recent collaborators. What if your former student graduated 20 years ago, however, and you have not collaborated since? What if one of the authors is a former collaborator with whom you have not published or spoken in 3 years? Few journals provide potential reviewers with specific instructions on exactly what constitutes a conflict of interest. Most rely on the judgment of the reviewers and ask them to err on the side of caution. *Maintaining the integrity of peer review thus requires individuals to avoid not only actual conflicts of interest but also possible conflicts of interest.* In other words, if the author of a manuscript might believe you are biased, positively or negatively, you should not review the paper even if you are confident you can provide an objective review.

You may find it difficult to evaluate objectively the work of a colleague you dislike; most would agree that this constitutes a bias. A more interesting question is whether you should review a manuscript written by a colleague whose work you consider to be generally of poor quality. On one hand, your responsibility as a reviewer is to help the editor evaluate the quality of work submitted for publication and to ensure publication of only high quality research; if you know the work of a particular group well, you may be uniquely qualified to explain why a particular manuscript does not merit publication. Yet, if you consider a particular group's work to be poor, it is likely that the authors will consider you misinformed or even biased against them. Fortunately, this is another situation where multiple reviewers provide a safeguard. If you explain carefully why you believe the work is flawed, the editor can balance your evaluation against those of other reviewers in making a final decision.

This discussion raises another issue that is critical to consider, especially by novice reviewers. The primary responsibility of the reviewer is to recommend whether to accept or reject a particular manuscript; in most cases, however, an intermediate step is to recommend potential revisions that would improve the work. By explaining carefully the strengths and weaknesses of a manuscript, you not only help the authors improve the manuscript, you also help advance their research and the field in general. Few things are more frustrating to an author than receiving a review that rates the paper poorly without providing specific criticisms. By contrast, few things are more helpful than a critique that suggests a critical new experiment or an alternative interpretation of your data that you had not considered previously. Again, however, we see that the reviewer is faced with an ethical decision. What if a reviewer realizes that the authors have missed the most important experiment or calculation? This reviewer could argue that the paper should be rejected because the authors have not proved their hypothesis or provided convincing results and then pursue the correct approach on his or her own and publish it later; similarly, this reviewer could allow the paper to be published and pursue and publish the correct result while referring to the paper and its error. Alternatively, the reviewer could provide significant guidance to the authors so that they can pursue the correct approach and publish what would be an important result for which they may be congratulated without any recognition of your anonymous contribution. Who would you ask for advice in this situation?

To guide your decision, it is critical to remember that new ideas, methods, and results contained within unpublished work that is undergoing peer review is privileged information. The editors and reviewers must treat the information confidentially and not use it for personal gain; you as a reviewer must do the same. That is, while unpublished, everyone must respect ownership of the information by the authors. Of course, with increased promotion of transparency and open access of information, many papers are now submitted without peer review to on-line archives. These archives not only increase earlier accessibility to the information, they document priority and thereby avoid some of the risks associated with peer review. Being freely available, however, mistakes discovered therein can now be used by anyone to correct the errors and receive credit for the new findings. Indeed, mistakes contained therein that may have been caught by a reviewer and corrected by you will now be archived. As in most cases, therefore, there are advantages and disadvantages to archiving ideas prior to peer review.

Exercise 8.3

Imagine that you are a third-year Ph.D. student and your faculty advisor asks you to review a paper that he or she was asked to review by a top journal. If you have never reviewed a paper before and have not yet published a paper, how should you proceed? What information would you expect your advisor to provide? If you complete the review and it is to be submitted to the journal based solely on your evaluation, should the editor be so notified? Should you get the "credit" for the review? How do you think the authors would feel if they disagreed with the conclusions of the review, if very negative, and they learned that it was conducted by a student?

8.2.2 Submitted Proposals

As with peer review of papers, one should always avoid conflicts of interest when asked to peer review proposals submitted for consideration for possible funding. Determining whether to excuse yourself from reviewing a particular grant is often simpler than for a particular journal article because funding agencies usually issue more specific guidelines than journals. At the NIH, for example (see https://grants.nih.gov/policy/peer/peer-coi.htm), you may not participate in the review of any grant application from your institution; if you are on the panel, you must leave the room during the discussion of these applications. You must also excuse yourself from the review of applications involving any colleague with whom you have published during the past 3 years. Furthermore, the review panel on which you sit may not review any application involving you; these grants must be sent to a different panel for consideration. NIH reviewers and staff work to identify and avoid other conflicts, whether real or perceived, not governed by specific guidelines. As a potential reviewer, it is your responsibility to disclose all potential conflicts of interest.

Years ago, De Vries et al. (2006) met with focus groups consisting of researchers and formulated a survey that they eventually conducted and reported in the *Nature* article "Scientists Behaving Badly" (Martinson et al., 2005), which was discussed earlier in this book. In the focus groups, they found researchers to be less worried about frank plagiarism or fabrication of data than about handling "fuzzier" situations that arise in science and engineering, such as properly apportioning credit for or ensuring the protection of ideas and discoveries. In particular, the primary concern of the researchers interviewed was potential theft of their ideas during the review of proposals. One commented (De Vries et al., 2006).

> I'm always wary of submitting grants to [NIH] study sections, because those people who sit on the study sections, it's not unknown for them to take your ideas, kill your grant, and then take and do it. And I think all of us have either had that happen to them or know somebody who had that happen to them.

That is, a challenging issue that arises during the review of proposals is deciding how to handle information contained within the applications. Although each application is confidential and must be deleted following review, a reviewer will necessarily remember many of the things that were read (and heard during a panel review). Indeed, in addition to the importance of fulfilling a professional responsibility, reviewing proposals often benefits a reviewer in three ways: you see the difference between well-prepared and poorly prepared applications, which can help you to prepare more competitive applications, you learn more about the overall process and what other reviewers value, and you are exposed to things that you may not have been aware of, including important references, new instrumentation or materials, useful experimental methods or computational tools, and so forth. Indeed, you may even learn of individuals who would be good potential collaborators. Yet, because applications are confidential, what information can you use and when? A good rule of thumb is that *any information available in the public domain can be used in good conscience*, including published papers, commercially available instruments, materials, software, and contacts listed on the Web for individuals in academia and industry. In other words, if you can obtain the same information elsewhere, it can be used.

By contrast, novel ideas (e.g., new experimental protocols or methods to solve a complex equation) that are not available in the public domain should not be used. Some might ask in this regard if it would be acceptable to contact the investigators and ask for permission to use their ideas. Albeit perhaps surprising, the answer is that you are not supposed to discuss any aspect of the grant application or its review with the applicant(s), hence it is not appropriate to seek such permissions. Rather, one should wait for the ideas to be made public by the investigators, often via a conference presentation, published paper, or posting on the Web. Once such disclosure has been made, it is then acceptable to use the available information or to contact the investigators for further clarification or possible collaboration. If the disclosure appears via a patent rather than via free information, however, one must then respect the conditions of the patent.

That it is appropriate to use information in the public domain should be obvious, but this does not resolve all potential issues. What if the applicants did not get funded and they were unable to pursue their ideas? What if they were funded but decided to pursue a different approach? In other words, is it good for science and engineering in general to let an excellent idea die simply because those who conceived the idea could not bring it to fruition? Is there a statute of limitations on such ideas or should there be? What if you wish to apply their idea to a completely different problem, one they are most likely not interested in and would never pursue? What if you were already planning to do the same experiment or one related closely—do you need to forego your experiment to avoid what may appear to be misconduct? Who should you approach to find answers to such questions?

Exercise 8.4

We identified some situations regarding "intellectual property" that may arise when a person reviews a proposal in confidence. Another situation could arise if you submitted a grant application to the NIH that needed to be revised and resubmitted, delaying the start of the project. What could you do if you suspected that someone on the review panel was purposely trying to delay your research so that they could pursue the same idea? List five other potential situations that could arise in grant reviews and discuss potential ethical issues.

8.2.3 Professional Conferences

One of the primary mechanisms by which science and engineering advance is the free sharing of ideas with others, most prominently via the archival journal article. This avenue of free exchange is safeguarded by the universal convention that one can use ideas within a published work if credit is given to the originator(s) of those ideas via appropriate citation (e.g., Smith et al., 2021). Another time-honored mechanism of free exchange of ideas is the professional conference where investigators ranging from seasoned researchers to beginning students gather together to present and discuss their ideas, findings, and concerns. When presenting such work (recall Chap. 5), it is both common and prudent to indicate on the slide or otherwise during the presentation the published work from which the information derives, if appropriate. In this way the listener can quickly take note of the source and later find associated details as well as provide appropriate attribution. Indeed, the oral presentation of published work is an excellent way to highlight, that is, draw attention to published work, and this serves an important function in advancing the field.

There are other cases, however, when the presented ideas, findings, or concerns are new, not previously published or patent protected. Indeed, some conferences encourage the presentation of new ideas. In this case, it is both common and prudent to denote on the slide or otherwise during the presentation that the work is "unpublished" or "confidential," as appropriate. This not only reveals that the information is new, but also signals to the listener that there is a need to respect the origin and thus ownership of the information given. By contrast to the situation of peer review of a paper or proposal, it is appropriate for any listener to approach a speaker to discuss the potential use of unpublished ideas, materials, or information—indeed, many useful collaborations have emerged in this way. As before, however, it is inappropriate to use unpublished or unprotected information specifically to "scoop" the presenter by publishing or patenting something that was presented in good faith. Effective communication, or discourse, with the presenter is the best way forward in this case. That said, one of the reasons that conferences and invited seminars are so common and useful is that one can learn of new applications and methods, which when used appropriately can drive new avenues of research.

Exercise 8.5

Read *The Double Helix* by James Watson and discuss how Watson and Crick handled unpublished information that was available to them.

8.3   Concluding Remarks

Recall from the Preface that it was not the goal of this book to be a standalone source on matters of style or ethics in communication. Rather, our goal was to motivate the reader to develop an effective, individual style of communicating and a personal commitment to integrity because it matters. We sought to raise questions, not answer them. Our best advice is simply to seek advice from good role models whose writing and ethics you respect—learn from others, solicit constructive feedback on oral presentations as well as drafts of manuscripts and proposals, and discuss potential concerns on ethical matters with peers, advisors, supervisors, and administrators. Work hard to ensure that, when you look back over your career, you will be proud of a job well done.

References

De Vries, R., Anderson, M. S., & Martinson, B. C. (2006). Normal misbehavior: Scientists talk about the ethics of research. *Journal of Empirical Research on Human Research Ethics, 1*, 43–50. https://doi.org/10.1525/jer.2006.1.1.43

Martinson, B. C., Anderson, M. S., & de Vries, R. (2005). Scientists behaving badly. *Nature, 435*, 737–738. https://doi.org/10.1038/435737a

Sampat, B. N. (2006). Patenting and U.S. academic research in the 20th century: The world before and after Bayh Dole. *Research Policy, 35*, 722–789.

2.3 Concluding Remarks

References

Index